实时高频GNSS地震监测与预警

李星星 著

Real-Time High-Rate GNSS
Techniques for
Earthquake Monitoring and Early Warning

武汉大学出版社

图书在版编目(CIP)数据

实时高频 GNSS 地震监测与预警 = Real-time High-rate GNSS Techniques for Earthquake Monitoring and Early Warning/李星星著. —武汉：武汉大学出版社,2017.12
ISBN 978-7-307-19837-1

Ⅰ.实… Ⅱ.李… Ⅲ.①地震监测—研究 ②地震预报—研究 Ⅳ.P315.7

中国版本图书馆 CIP 数据核字(2017)第 276564 号

责任编辑:谢群英　　责任校对:汪欣怡　　版式设计:汪冰滢

出版发行:**武汉大学出版社**　(430072　武昌　珞珈山)
（电子邮件：cbs22@whu.edu.cn　网址：www.wdp.com.cn）
印刷:虎彩印艺股份有限公司
开本:720×1000　1/16　印张:12　字数:222 千字　插页:1
版次:2017 年 12 月第 1 版　　2017 年 12 月第 1 次印刷
ISBN 978-7-307-19837-1　　定价:36.00 元

版权所有，不得翻印；凡购我社的图书，如有质量问题，请与当地图书销售部门联系调换。

Preface

In recent times increasing numbers of high-rate GNSS stations have been installed around the world and set-up to provide data in real-time. These networks provide a great opportunity to quickly capture surface displacements, which makes them important as potential constituents of earthquake/tsunami monitoring and warning systems. The appropriate GPS real-time data analysis with sufficient accuracy for this purpose is a main focus of the current GNSS research. The objective of this book is to develop high-precision GNSS algorithms for better seismological applications. The core research and the contributions of this book are summarized as following:

With the availability of real-time high-rate GNSS observations and precise satellite orbit and clock products, the interest in the real-time Precise Point Positioning (PPP) technique has greatly increased to construct displacement waveforms and to invert for source parameters of earthquakes in real time. Furthermore, PPP ambiguity resolution approaches, developed in the recent years, overcome the accuracy limitation of the standard PPP float solution and achieve comparable accuracy with relative positioning. In this book, we introduce the real-time PPP service system and the key techniques for real-time PPP ambiguity resolution. We assess the performance of the ambiguity-fixed PPP in real-time scenarios and confirm that positioning accuracy in terms of root mean square (RMS) of 1.0 cm-1.5 cm can be achieved in horizontal components. For the 2011 Tohoku-Oki (Japan) and the 2010 El Mayor-Cucapah (Mexico) earthquakes, the displacement waveforms, estimated from ambiguity-fixed PPP and those provided by the accelerometer instrumentation are consistent in the dynamic component within few centimeters. The PPP fixed solution not only can improve the accuracy of coseismic displacements, but also provides a reliable recovery of earthquake magnitude and of the fault slip distribution in real time.

We propose an augmented point positioning method for GPS based hazard monitoring, which can achieve fast or even instantaneous precise positioning without relying on data of a specific reference station. The proposed method overcomes the limitations of the currently mostly used GPS processing approaches of relative positioning and global precise point positioning. The advantages of the proposed approach are demonstrated by using GPS data, which was recorded during the 2011 Tohoku-Oki earthquake in Japan.

We propose a new approach to quickly capture coseismic displacements with a single GNSS receiver in real-time. The new approach can overcome the convergence problem of precise point positioning (PPP), and also avoids the integration process of the variometric approach. Using the results of the 2011 Tohoku-Oki earthquake, it is demonstrated that the proposed method can provide accurate displacement waveforms and permanent coseismic offsets at an accuracy of few centimeters, and can also reliably recover the moment magnitude and fault slip distribution. We investigate three current existing single-receiver approaches for real-time GNSS seismology, comparing their observation models for equivalence and assessing the impact of main error components. We propose some refinements to the variometric approach and especially consider compensating the geometry error component by using the accurate initial coordinates before the earthquake to eliminate the drift trend in the integrated coseismic displacements.

We propose an approach for tightly integrating GPS and strong motion data at raw observation level to increase the quality of the derived displacements. The performance of the proposed approach is demonstrated using 5 Hz high-rate GPS and 200 Hz strong motion data collected during the El Mayor-Cucapah earthquake (Mw 7.2, 4 April, 2010) in Baja California, Mexico. The new approach not only takes advantages of both GPS and strong motion sensors, but also improves the reliability of the displacement by enhancing GPS integer-cycle phase ambiguity resolution, which is very critical for deriving displacements with highest quality. We also explore the use of collocated GPS and seismic sensors for earthquake monitoring and early warning. The GPS and seismic data collected during the 2011 Tohoku-Oki (Japan) and the 2010 El Mayor-Cucapah (Mexico) earthquakes are analyzed by using a tightly-coupled integration. The performance of the integrated results are validated by both time and frequency domain analysis. We detect the P-wave arrival and observe small-scale features of the movement from the integrated results and locate the epicenter. Meanwhile, permanent offsets are extracted from the integrated displacements highly accurately and used for reliable fault slip inversion and magnitude estimation.

Contents

List of Abbreviations ··· 1

List of Related Publications ··· 1

Chapter 1 Introduction ··· 1

Chapter 2 High-Rate GNSS Seismology Using Real-Time PPP with Ambiguity Resolution ·············· 9

2.1　Introduction ·· 9
2.2　Real-Time PPP System and Algorithms ················ 11
　2.2.1　*Real-Time PPP Service System* ······················· 11
　2.2.2　*Observation Model* ·· 14
　2.2.3　*Real-Time Precise Point Positioning* ··············· 17
　2.2.4　*Real-Time Estimation of the Uncalibrated Phase Delay* ·· 18
　2.2.5　*Ambiguity Resolution for Precise Point Positioning* ·· 20
2.3　Accuracy Assessment in Real-Time Scenarios ············ 21
2.4　Application to the 2011 Tohoku-Oki Earthquake ········ 26
　2.4.1　*GPS Data and Analysis* ····································· 26
　2.4.2　*Comparing GPS and Seismic Waveforms* ········ 29
　2.4.3　*Fault Slip Inversion* ··· 35

1

2.5 Application to the 2010 El Mayor-Cucapah Earthquake ········ 38
   2.5.1 *GPS Data and Analysis* ········ 38
   2.5.2 *Comparing GPS and Seismic Waveforms* ········ 41
   2.5.3 *Fault Slip Inversion* ········ 43
2.6 Conclusions ········ 47

Chapter 3 Augmented PPP for Seismological Applications Using Dense GNSS Networks ········ 50
3.1 Introduction ········ 50
3.2 Augmented PPP Approach ········ 52
3.3 Application of Augmented PPP Approach and Results ········ 56
3.4 Conclusions ········ 70

Chapter 4 Temporal Point Positioning Approach for GNSS Seismology Using a Single Receiver ········ 72
4.1 Introduction ········ 72
4.2 Temporal Point Positioning Approach ········ 74
4.3 Application of TPP Approach and Results ········ 78
4.4 Single-Receiver Approaches for Real-Time GNSS Seismology ········ 87
   4.4.1 *Comparison of Analysis Methods* ········ 87
   4.4.2 *Error Analysis and Precision Validations* ········ 96
   4.4.3 *Application to the 2011 Tohoku-Oki Earthquake* ········ 109
4.5 Conclusions ········ 117

Chapter 5 Tightly-Integrated Processing of Raw GNSS and Accelerometer Data ········ 120
5.1 Introduction ········ 120

5.2 Overview of Combining GPS and Accelerometer Data 121
5.3 The Tightly-Integrated Algorithm 123
5.4 Results 127
    5.4.1 *Comparison of GPS, Seismic and Integrated Waveforms* 128
    5.4.2 *Detection of P-Wave Arrival* 147
    5.4.3 *Extraction of Permanent Offset and Fault Slip Inversion* 152
5.5 Conclusions 155

Chapter 6 Conclusions and Outlook 157

References 162

Acknowledgments 175

List of Abbreviations

AC	Analysis Center
ARP	Antenna Reference Point
BDS	the Chinese BeiDou Navigation Satellite System
BKG	Federal Agency for Cartography and Geodesy, Germany
C/A	Coarse/Acquisition Code
CDDIS	Crustal Dynamics Data Information System
CODE	Centre of Orbit Determination in Europe
CORS	Continuously Operating Reference Stations
DCB	Differential Code Biases
DD	Double Difference
DF	Dual-Frequency
DFG	Deutsche Forschungs Gemeinchaft (i.e. German Research Foundation)
DOD	the U.S. Department of Defense
ECMWF	European Centre for Medium-Range Weather Forecasts
ESA	European Space Agency
EU	European Union
EEW	Earthquake Early Warning
Galileo	the European Union Satellite Navigation System
GEO	Satellites in Geostationary Orbit
GFZ	Helmholtz-Centre Potsdam-GFZ German Research Centre for Geosciences

List of Abbreviations

GIM	Global Ionospheric Map
GLONASS	the Russian GLOBAL Navigation Satellite System
GLOT	GLONASS Time
GMF	Global Mapping Function
GNSS	Global Navigation Satellite System
GPS	Global Positioning System
GPST	GPS Time
GSI	Geospatial Information Authority
IAR	Integer Ambiguity Resolution
IGR	IGS Rapid Orbit
IGS	International GNSS Service
IGSO	Inclined Geosynchronous Orbit
IGU	IGS Ultra-Rapid Orbit
IOV	In-Orbit Validation
IPP	Ionospheric Pierce Point
ITRF	International Terrestrial Reference Frame
LC	Ionosphere-Free Linear Combination
LEO	Low Earth Orbit
MEO	Medium Altitude Earth Orbit
MET	Meteorology
MW_WL	MW Widelane Linear Combination
NASA	National Aeronautics and Space Administration
NRTK	Network-Based Real-Time Kinematic Positioning
OMC	Observation Minus Computation
PCO	Phase Centre Offsets
PCV	Phase Centre Variation
PNT	Positioning, Navigation and Timing
PPP	Precise Point Positioning
PPP-RA	Precise Point Positioning Regional Augmentation
RMS	Root Mean Square
RTK	Real-Time Kinematics

SAPOS	Satellite Positioning Service of the German State Survey
SD	Single Difference
SP3	IGS Standard Product 3
SPS	Standard Positioning Service
STD	Slant Total Delay
UD	Un-Differenced
UPD	Un-Calibrated Phase Delays
UTC	Coordinated Universal Time
WL	Widelane Combination
WGS-84	Word Geodetic System 1984
ZHD	Zenith Hydrostatic Delay
ZTD	Zenith Total Delay
ZWD	Zenith Wet Delay

List of Related Publications

1. Li, X., M. Ge, X. Zhang, Y. Zhang, B. Guo, R. Wang, J. Klotz, and J. Wickert (2013). Real-time high-rate coseismic displacement from ambiguity-fixed precise point positioning: Application to earthquake early warning. Geophys. Res. Lett., 40(2), 295-300, doi:10.1002/grl.50138.

2. Li, X., M. Ge, Y. Zhang, R. Wang, P. Xu, J. Wickert, and H. Schuh (2013). New approach for earthquake/tsunami monitoring using dense GPS networks. Sci. Rep., 3, 2682, doi:10.1038/srep02682.

3. Li, X., M. Ge, B. Guo, J. Wickert, and H. Schuh (2013). Temporal point positioning approach for real-time GNSS seismology using a single receiver. Geophys. Res. Lett., 40(21), 5677-5682, doi:10.1002/2013GL057818.

4. Li, X., M. Ge, Y. Zhang, R. Wang, B. Guo, J. Klotz, J. Wickert, and H. Schuh (2013). High-rate coseismic displacements from tightly integrated processing of raw GPS and accelerometer data. Geophys. J. Int.

5. Li, X., X. Zhang, and B. Guo (2013). Application of collocated GPS and seismic sensors to earthquake monitoring and early warning. Sensors, 13:14261-14276.

List of Related Publications

6. Li, X., M. Ge, C. Lu, Y. Zhang, R. Wang, J. Wickert, and H. Schuh (2014). High-rate GPS seismology using real-time precise point positioning with ambiguity resolution. IEEE transactions on geoscience and remote sensing, pp.1-15.

7. Li, X., B. Guo, C. Lu, M. Ge, J. Wickert, and H. Schuh (2014). Real-time GNSS seismology using a single receiver. Geophys. J. Int. doi: 10.1093/gji/ggu113.

Chapter 1 Introduction

Recent destructive earthquakes that struck Sumatra, Indonesia (Mw 9.2) in 2004, Wenchuan, China (Mw 7.9) in 2008, Maule, Chile (Mw 8.8) in 2010 and Tohoku, Japan (Mw 9.0) in 2011 have once again brought us to focus the urgent need for earthquake monitoring and early warning. Rapid source and rupture inversion for large earthquakes is critical for seismic and tsunamigenic hazard mitigation (Allen and Ziv, 2011; Ohta et al., 2012), and earthquake-induced site displacement is key information for such source and rupture inversions. For earthquake early warning (EEW) systems, the estimation of accurate coseismic displacements and waveforms is needed in real-time. Traditionally, displacements are obtained by double integration of observed accelerometer signals or single integration of velocities observed with broadband seismometers (Kanamori, 2007; Espinosa-Aranda et al., 1995; Allen and Kanamori, 2003). The broadband seismometers are likely to clip the signal in case of large earthquakes. Although strong-motion accelerometer instruments do not clip, the displacement converted from acceleration could be degraded significantly by drifts caused by tilts and the non-linear behavior of the accelerometer (Trifunac and Todorovska, 2001; Boore, 2001).

Since Remondi (1985) first demonstrated cm-level accuracy of kinematic GPS, Hirahara et al. (1994) labeled kinematic GPS as GPS seismology, which has since attracted more and more attention and applications in seismology (see, e.g., Ge et al. 2000; Larson et al., 2003). High-rate GNSS (e.g., 1 Hz or higher frequency) measures displacements directly and can provide reliable estimates of broadband displacements, including static offsets and dynamic motions of arbitrarily large magnitude (Larson et al., 2003; Bock et al., 2004). GPS-based seismic source characterization has been demonstrated in near-and far-field with remarkable results (Nikolaidis et al., 2001; Larson et al., 2003; Bock et al., 2004; Ohta et al., 2008; Yokota et al., 2009; Avallone et al., 2011; Melgar et al., 2012; Crowell et al., 2012). GNSS-derived displacements can be used to quickly estimate earthquake magnitude, model finite fault slip, and also play an important role in earthquake/tsunami early warning (Blewitt et al., 2006; Wright et al., 2012; Hoechner et al., 2013). Consequently in the recent years, dense GPS monitoring networks have been built in seismically active regions, e.g., Japan's GEONET (the GPS Earth Observation Network System, http://www.gsi.go.jp/) and UNAVCO's Plate Boundary Observatory (PBO, http://pbo.unavco.org/). These networks are complementary to seismic monitoring networks and contribute significantly to earthquake/tsunami early warning and hazard risk mitigation (Blewitt et al., 2006; Crowell et al., 2009).

Currently, two processing strategies are mainly used in most of the studies related to GPS seismology and tsunami warning: relative baseline/network positioning (e.g., Nikolaidis et al., 2001; Larson et al., 2003; Bock et al., 2004, Blewitt et al., 2006) and precise point positioning (PPP) (Zumberge et al., 1997). For relative kinematic positioning, at least one nearby reference station should be used for removing most of biases and recovering integer feature of ambiguity parameters by forming double-differenced ambiguities. Consequently, ambiguities can always be fixed to integers even instantaneously for achieving high positioning accuracy of few cm (Bock et al., 2011; Ohta et al., 2012). Therefore, it is already applied in real-time displacement monitoring (e.g., Crowell et al., 2009). The technique of instantaneous positioning (Bock et al., 2000) is one typical real-time relative positioning method and is integrated into EEW system (Crowell et al., 2009) and is demonstrated by applying the result for centroid moment tensors (CMT) computation (Melgar et al., 2012), finite fault slip inversion (Crowell et al., 2012) and P-wave detection by combining collocated accelerometer data and the GPS displacements using a Kalman filter (Bock et al., 2011; Tu et al., 2014). The real-time kinematic (RTK) technique is also utilized by Ohta et al. (2012) to analyze the displacement of the 2011 Tohoku-Oki earthquake. All of the previously mentioned studies used the relative positioning technique, which is able to guarantee a high accuracy at 1 cm level. However, for the relative positioning technique, GPS data from a network is analyzed simultaneously to estimate station positions. It is complicated by the need to assign baselines, overlapping Delaunay triangles, or overlapping sub-networks. This is a significant limitation for the challenging simultaneous and precise real-time analysis of GPS data from hundreds or thousands of ground stations. Furthermore, intermittent station dropouts complicate the network-based relative positioning. Relative positioning also requires a local reference station, which might itself be displaced during a large seismic event, resulting in misleading GPS analysis results. In the case of large earthquakes, such as the Mw 9.0 Tohoku-Oki event in Japan, the reference station may also be significantly displaced, even when it is several hundred kilometers away from the event. The reference station should be sufficiently far from the focal region, but must also be part of a sub-network that has relatively short baselines (usually within several tens of kilometers). As the baseline length increases, the accuracy of relative positioning would be significantly reduced because the atmospheric effects and satellite ephemeris errors become less common and thus cannot be effectively cancelled out by double-difference technique.

PPP provides a new concept of positioning service by using precise orbit and clock products generated from a global reference network (Kouba and Héroux, 2001). Kouba (2003) demonstrated that PPP using the orbit and clock products of the International GNSS Service (IGS) can be used to detect seismic waves and satisfy the requirement of the GPS seismology. Wright et al. (2012) used PPP in real-time mode with broadcast clock and orbital corrections to give station positions every 1 sec and then carry out a simple static inversion to determine the portion of the fault that slipped and the earthquake magnitude. The PPP technique can provide "absolute" coseismic displacements with respect to a global reference frame (defined by the satellite orbits and clocks) with a stand-alone GPS receiver. A PPP processing system uses information from a global reference network, which is applied to the monitoring stations, consequently the derived positions are referred to the global network, which itself is hardly affected by the earthquake displacement.

Thanks to the development of JPL's Global Differential GPS System (GDGPS) and the International GNSS Service (IGS) real-time pilot project (RTPP), real-time precise satellite orbit and clock products are now available online and PPP is widely recognized as a promising positioning technique (Caissy 2006; Dow et al., 2009; Bar-Sever et al., 2009). However, standard PPP (float ambiguity) has limited accuracy in real-time applications because of unresolved integer-cycle phase ambiguities. PPP ambiguity resolution developed in recent years provides an important promise to achieve comparable accuracy with relative/ network positioning (Ge et al., 2008; Li et al., 2013a). The German Research Center for Geosciences (GFZ), as one of the IGS data analysis centers, is operationally providing GPS orbits and clocks, uncalibrated phase delays (UPDs) and differential code biases (DCBs) for real-time PPP service with ambiguity-fixing (Ge et al., 2011; Li 2012). The performance is further enhanced by new algorithms for speeding up the reconvergence through estimation of ionospheric delays (e.g., Geng et al., 2010; Zhang and Li, 2012; Li et al., 2013a). Although a convergence period of about 20 minutes is still required, PPP is able to achieve cm-level positioning accuracy in real-time without the need for dedicated reference stations (Li et al., 2013b).

Colosimo et al. (2011) proposed a variometric approach to determine the change of position between two adjacent epochs (namely delta position) based upon the time single-differences of the carrier phase observations, and then displacements of the station are obtained by a single integration of the delta positions. This approach does not suffer from convergence process, but the single integration from delta positions to displacements is accompanied by a drift due to the potential uncompensated errors. Usually, a limited duration of 3-4 minutes may be enough for large displacements retrieving. Under the assumption that the variometric-based displacement has a linear trend within few minutes, the estimated displacements after linear trend removal are demonstrated to be at a level of a few centimeters (Branzanti et al., 2013).

Recent advances in the performance of real-time high-rate GPS, estimates of permanent displacement directly, mean that its use can potentially be complementary to the seismic-based methodologies for earthquake early warning. The main weaknesses of current GPS measurements are the lower sampling rates (1-50Hz) and the larger high-frequency noise contribution, and so the GPS-derived dynamic motions are not accurate enough to identify the first arrival wave (P-wave) with only millimeter-level amplitude. The noise of GPS displacements is basically white across the whole seismic frequency band. While strong motion sensors are able to sample at very high rates (e.g. 200Hz) and perform very well in the high-frequency range as it is much more sensitive to ground motions than GPS receiver, especially in the vertical direction. However, the acceleration is accompanied by unphysical drifts due to sensor rotation and tilt (Trifunac and Todorovska, 2001; Lee and Trifunac, 2009), hysteresis (Shakal and Petersen, 2001), and imprecision in the numerical integration process (Boore et al., 2002; Smyth and Wu, 2006). Its noise level, viewed in terms of displacement, will rise with decreasing frequency: at some frequency this noise level will exceed that of GPS. Therefore, GPS and seismic instruments can be mutually beneficial for seismological applications because weaknesses of one observation technique are offset by strengths in the other.

The complementary nature of GPS and seismic sensors for station displacement estimation and P-wave detection are well recognized and the integrated processing of the two dataset is a hot topic in GPS seismology for obtaining more accurate and reliable displacements and P-wave arrival time. Several loosely-integrated approaches have been proposed to fuse accelerometer with collocated GPS displacement data (Emore et al., 2007; Bock et al., 2011). As the GPS coordinates are already estimated prior to integration with the accelerometer, the precise dynamic information provided by accelerometers cannot be used to enhance the GPS-only solutions in these integration algorithms.

Currently, the available real-time high-rate GNSS data streams have the potential to contribute to EEW system. GFZ has been working on the real-time precise positioning for geohazard monitoring for years. The EPOS-RT software has been developed for providing worldwide real-time PPP service. The accuracy and reliability of the global PPP should be improved for better monitoring of geohazard, especially for those causing rather small displacement. It is a big challenge to achieve precise and reliable displacements in real-time. Thus this thesis will focus on the development of high-precision GNSS algorithms for better seismological applications. We also integrate the accelerometer data into the GNSS data processing in order to combine all the advantages of GPS and seismic sensors. This thesis includes the following chapters:

Firstly, Chapter 1 presents the motivation, background and research objectives of this book and specifies the contributions of this research.

In the Chapter 2, the ambiguity-fixed PPP method is developed to estimate high-rate coseismic displacement in real-time. This ambiguity-fixed PPP algorithm is described in detail. The PPP displacement waveforms are analyzed and compared with seismic waveforms and the application of PPP waveforms to EEW is described.

Chapter 3 proposes a novel method for fast or even instantaneous positioning, making full use of the currently available global PPP service and regional GPS monitoring networks. The derived atmospheric corrections at the stations with fixed ambiguities then can be provided to other monitoring stations for instantaneous ambiguity resolution, so that precise displacements can always be achieved within a few seconds. The new method does not depend on a specific reference station and therefore the analysis results will not be affected by simultaneous shaking of any particular station. It also has better flexibility and efficiency compared to complicated network/subnetwork analysis.

Chapter 4 proposes a new approach for estimating coseismic displacements with a single receiver in real-time. The approach overcomes not only the disadvantages of the PPP and RP techniques, but also decreases the described drift in the displacements derived from the variometric approach. The coseismic displacement could be estimated with few centimeters accuracy using GNSS data around the earthquake period. The efficiency of the new approach is validated using 1 Hz GNSS data, collected during the Tohoku-Oki earthquake (Mw 9.0, 11 March, 2011) in Japan.

Chapter 5 proposes an approach of integrating the accelerometer data into the GPS data processing in order to take full use of the complementary of GPS and seismic sensors. Instead of combing the GPS-derived displacements with the accelerometer data, a tightly-integrated filter is developed to estimate seismic displacements from GPS phase and range and accelerometer observations. The performance of the proposed tightly-integrated approach was validated by the 2010, Mw 7.2 El Mayor-Cucapah earthquake (Mw 7.2, 4 April, 2010) in Baja California, Mexico and the Tohoku-Oki earthquake (Mw 9.0, 11 March, 2011) in Japan.

Finally, Chapter 6 summarizes the main results as obtained in the previous chapters, presents the final conclusions and suggests recommendations for the future work.

Chapter 2 High-Rate GNSS Seismology Using Real-Time PPP with Ambiguity Resolution

2.1 Introduction

Earthquake-induced site displacement is key information for locating the epicentre and estimating the magnitude of earthquakes. For earthquake early warning (EEW) systems, the estimation of accurate coseismic displacements and waveforms is needed in real-time. Traditionally, displacements are obtained by double integration of observed accelerometer signals or single integration of velocities observed with broadband seismometers (Kanamori, 2007; Espinosa-Aranda et al., 1995; Allen and Kanamori, 2003). Compared to seismic instruments that are subject to drifts (accelerometer) or that clip the signal in case of large earthquakes (broadband seismometer), GPS receiver observes displacements directly, making it particularly valuable in case of large earthquakes (Larson et al., 2003; Blewitt et al., 2006).

Until now, post-processing or simulated real-time relative kinematic positioning is utilized in most of the studies related to GPS seismology (e.g., Nikolaidis et al., 2001; Larson et al., 2003; Bock et al., 2004) and tsunami warning (e.g., Blewitt et al., 2006). For relative kinematic positioning, at least one nearby reference station should be used for removing most of biases and recovering integer feature of ambiguity parameters by forming double-differenced ambiguities. Consequently, ambiguities can always be fixed to integers even instantaneously for achieving high positioning accuracy of few cm (Bock et al., 2011; Ohta et al., 2012). However, the relative positioning technique relies on reference stations located closely. To monitor large regions, consequently a large number of reference stations is required which increases costs and system complexity. More critical is the fact that the derived displacement also depends on the earthquake-induced movement of the reference stations.

Thanks to the development of JPL's Global Differential GPS System (GDGPS) and the International GNSS Service (IGS) real-time pilot project (RTPP), real-time precise satellite orbit and clock products are now available online and PPP is widely recognized as a promising positioning technique (Caissy 2006; Dow et al., 2009; Bar-Sever et al., 2009). However, standard PPP (float ambiguity) has limited accuracy in real-time applications because of unresolved integer-cycle phase ambiguities. In recent years, integer ambiguity fixing approach for PPP has been developed to improve its performance (Ge et al., 2008; Li et al., 2013).

In this chapter, we introduce the system structure of the real-time PPP service by taking the operational real-time PPP service, developed at the German Research Center for Geosciences (GFZ) for both precise positioning and geophysical applications as an example. The GPS data processing strategies including the observation model, real-time PPP, and PPP ambiguity resolution are described in detail. The positioning precision is carefully assessed in real-time scenarios. We estimate high-rate coseismic displacements in simulated real-time PPP mode for 1 Hz GPS data, collected during the Tohoku-Oki earthquake (Mw 9.0, March 11, 2011) in Japan and 5 Hz GPS data, collected during El Mayor-Cucapah earthquake (Mw 7.2, April 4, 2010) in Mexico. The derived displacement waveforms are analyzed, compared with seismic waveforms, and finally applied for fault slip inversion to further validate the real-time PPP service with ambiguity resolution.

2.2 Real-Time PPP System and Algorithms

2.2.1 Real-Time PPP Service System

In a global real-time PPP service system, real-time data from a global reference network with a certain number of evenly distributed stations is essential for generating precise orbits and clocks for precise positioning at the desired location. The real-time orbit is usually predicted based on orbits determined in a batch-processing mode using the latest available observations (generally 5 min sampling interval) due to the dynamic stability of the satellite movement. More challenging is the estimation of the satellite clock corrections, which must be updated much more frequently due to their short-term fluctuations (Zhang et al., 2011).

Under the framework of the IGS RTPP, data from a global real-time network of more than 100 stations is available and the related data communication for the observation, retrieving and product casting was established (Caissy 2006). Furthermore, several real-time analysis centers (RTAC) were established and have been running operationally to contribute their real-time products for comparison and combination. Most of the RTACs use a processing procedure similar to that for generating the IGS ultra-rapid orbits, but the update time of 6 hours could be shortened for a better orbit accuracy. The satellite clock products are estimated every 5 seconds together with receiver clock, ambiguity and zenith tropospheric delay (ZTD) parameters with fixed or tightly constrained satellite orbits and station coordinates. As in both the orbit determination and clock estimation, station coordinates are fixed, observations affected by earthquake-induced displacements will have large residuals and be rejected as outliers. As the displacement will not occur at the same time for most of the stations, precise satellite orbits and clocks are still available during the earthquake time.

With the real-time orbits and clocks, standard PPP can be carried out at the user-end where ionospheric delays are usually eliminated using the ionosphere-free combination and ZTD must be estimated as unknown parameter. Usually, it needs an initialization time of about 30 minutes to obtain position of cm-level accuracy. Initialization must be repeated if most of the satellites lose lock which is referred as to re-convergence and might occur during earthquakes. Moreover, as in relative positioning, the float solution could be further improved by integer ambiguity resolution. For recovering the integer feature of the ambiguities at the user-end, uncalibrated phase delay (UPD) at the satellites or similar products should be estimated and transmitted to users. The approach for this UPD estimation and integer ambiguity resolution at the user-end will be presented in detail in the following sections.

Taking the operational real-time PPP service, developed at GFZ, as example (Ge et al., 2011; Li et al., 2013), the data streams of about 80 globally distributed IGS stations are processed in real-time by using the EPOS-RT software for generating and broadcasting orbits and clocks for standard PPP service. Furthermore, for resolving the integer ambiguity at user-end, UPD corrections are generated and updated in real-time with the same data set which are used to estimate orbit and clock products. The data flow at the server-end, including precise GPS orbit determination, clock estimation and UPD retrieval is shown in Figure 2.1.

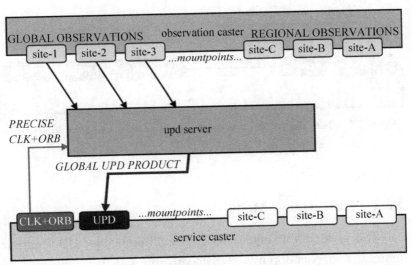

Figure 2.1 *Data flow at the server-end of a global real-time PPP service. The GPS clocks and orbits are basic products and UPD is for PPP ambiguity resolution at the user-end.*

At the user-end, the real-time product streams of orbits and clocks are received and applied to the observations for standard PPP. If UPD is also available, PPP with integer ambiguity resolution then can be performed. Due to the product transmission delay and higher sampling rate (e.g., 1Hz) at user-end, a linear extrapolation of a few seconds will be applied. The data flow at the user-end is shown in Figure 2.2.

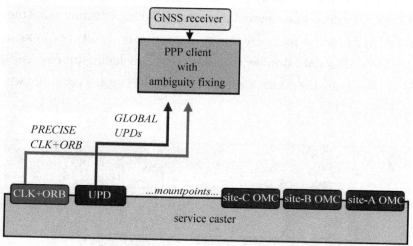

Figure 2.2   *Data flow at the user-end for standard PPP and PPP with integer ambiguity resolution.*

2.2.2 Observation Model

The observation equations for undifferenced (UD) carrier phase L and pseudorange P respectively, can be expressed as following:

$$L_{r,j}^s = \rho_{r\ g}^s - t^s + t_r + \lambda_j(b_{r,j} - b_j^s) + \lambda_j N_{r,j}^s - I_{r,j}^s + T_r^s + \varepsilon_{r,j}^s \quad (2.1)$$

$$P_{r,j}^s = \rho_{r\ g}^s - t^s + t_r + c(d_{r,j} + d_j^s) + I_{r,j}^s + T_r^s + e_{r,j}^s \quad (2.2)$$

where indices s, r, and j refer to the satellite, receiver, and carrier frequency, respectively; t^s and t_r are the clock biases of satellite and receiver; $N_{r,j}^s$ is the integer ambiguity; $b_{r,j}$ and b_j^s are the receiver-and satellite-dependent UPD; λ_j is the wavelength; $d_{r,j}$ and d_j^s are the code biases of the receiver and the satellite; $I_{r,j}^s$ is the ionospheric delay of the signal path at frequency j; T_r^s is the corresponding and frequency independent tropospheric delay; $e_{r,j}^s$ and $\varepsilon_{r,j}^s$ are measurement noise of the pseudorange and carrier phase observations. Furthermore, ρ_g denotes the geometric distance between the phase centers of the satellite and receiver antennas at the signal transmitting and receiving epoch, respectively. This means, that the phase center offsets and variations and station displacements by tidal loading must be considered. Phase wind-up and relativistic delays must also be corrected according to the existing models (Kouba and Héroux, 2001), although they are not included in the equations. A sidereal filter can be used to effectively reduce the impact of multipath error (Choi et al., 2004).

The slant tropospheric delay consists of the dry and wet components and both can be expressed by their individual zenith delay and mapping function. The tropospheric delay is usually corrected for its dry component with an a priori model, while the residual part of the tropospheric delay (considered as zenith wet delay Z_r) at the station r is estimated from the observations.

$$L_{r,j}^s = \rho_{r\ g}^s - t^s + t_r + \lambda_j (b_{r,j} - b_j^s) + \lambda_j N_{r,j}^s - I_{r,j}^s + m_r^s \cdot Z_r + \varepsilon_{r,j}^s \quad (2.3)$$

$$P_{r,j}^s = \rho_{r\ g}^s - t^s + t_r + c(d_{r,j} + d_j^s) + I_{r,j}^s + m_r^s \cdot Z_r + e_{r,j}^s \quad (2.4)$$

where m_r^s is the wet mapping function.

For multi-frequency observations, the ionospheric delays at different frequencies can be expressed as,

$$I_{r,j}^s = \kappa_j \cdot I_{r,1}^s; \quad \kappa_j = \lambda_j^2 / \lambda_1^2 \quad (2.5)$$

At the user-end, the real-time product streams of orbits and clocks are received and applied to the observations for standard PPP. If UPD is also available, PPP with integer ambiguity resolution then can be performed. Due to the product transmission delay and higher sampling rate (e.g., 1Hz) at user-end, a linear extrapolation of a few seconds will be applied. The data flow at the user-end is shown in Figure 2.2.

The ionospheric delays can be eliminated by forming the linear combination of observations at different frequencies. Usually, the ionosphere-free observation is employed in PPP (Kouba and Héroux, 2001). Alternatively, the dual-frequency data can be processed by estimating the slant ionospheric delays in raw observations as unknown parameters (Schaffrin and Bock 1988). The line-of-sight ionospheric delays on the L1 frequency $I^s_{r,1}$ are taken as parameters to be estimated for each satellite and at each epoch. In order to strengthen the solution, a priori knowledge about these delays can be utilized as constraints on the ionospheric parameters. The temporal change of the slant ionospheric delay of a satellite-station pair can be represented by a stochastic process according to their temporal correlation. Ionospheric gradient parameters could be used to take into account the spatial distribution of the ionospheric delay. The temporal correlation, spatial characteristics and ionospheric model constraints are comprehensively considered to speed up the convergence in PPP ambiguity resolution (Li et al., 2013c). These constraints, to be imposed on observations of a single station can be summarized as:

$$I^s_{r,t} - I^s_{r,t-1} = w_t, w_t \sim N(0, \sigma^2_{wt})$$
$$vI^s_r = I^s_r / f_{r,IPP} = a_0 + a_1 dL + a_2 dL^2 + a_3 dB + a_4 dB^2, \sigma^2_{vI} \quad (2.6)$$
$$I^s_r = \tilde{I}^s_r, \sigma^2_{\tilde{I}}$$

where t is the current epoch; $t-1$ is the previous epoch; w_t is a zero mean white noise with variance of σ^2_{wt} (generally a few millimeters for 1Hz sampling and an elevation-dependent weighting strategy is applied); vI^s_r is the vertical ionospheric delay with a variance of σ_{vI}^2; $f_{r,IPP}$ is the mapping function (Schaer et al., 1999) at the ionospheric pierce point (IPP); the coefficients $a_i (i = 0,1,2,3,4)$ describe the planar trend, the average value of the ionospheric delay over the station; a_1, a_2, a_3, and a_4 are the coefficients of the two second-order polynomials that used to fit the east-west and south-north horizontal gradients; dL and dB are the longitude and latitude difference between the IPP and the station location; $\tilde{I}_r^s$ is the ionospheric delay obtained from external ionospheric model with a variance of σ_I^2.

2.2.3 Real-Time Precise Point Positioning

Instead of the ionosphere-free linear combination, we use the raw carrier phase and pseudorange observations of Eqs. (2.3), (2.4), and (2.5) and estimate the slant ionospheric delay as unknown parameters with ionospheric constraint of (2.6). The linearized equations for (2.3) and (2.4) can be respectively expressed as following,

$$l_{r,j}^s = -\mathbf{u}_r^s \cdot \Delta \mathbf{r} - t^s + t_r + \lambda_j(b_{r,j} - b_j^s) + \lambda_j N_{r,j}^s - \kappa_j \cdot I_{r,1}^s + m_r^s \cdot Z_r + \varepsilon_{r,j}^s \tag{2.7}$$

$$p_{r,j}^s = -\mathbf{u}_r^s \cdot \Delta \mathbf{r} - t^s + t_r + c(d_{r,j} + d_j^s) + \kappa_j \cdot I_{r,1}^s + m_r^s \cdot Z_r + e_{r,j}^s \tag{2.8}$$

where $l_{r,j}^s$ and $p_{r,j}^s$ denote "observed minus computed" phase and pseudorange observables from satellite s to receiver r at the frequency j; $\mathbf{u}_r^s$ is the unit vector of the direction from receiver to satellite; $\Delta \mathbf{r}$ denotes the vector of the receiver position increment.

With the received real-time corrections of GPS satellite orbits, clocks and differential code biases (DCBs), the corresponding terms in the observation equations can be removed. The raw observation equations then can be simplified as,

$$l_{r,j}^s = -\mathbf{u}_r^s \cdot \Delta \mathbf{r} + t_r + \lambda_j \cdot B_{r,j}^s - \kappa_j \cdot I_{r,1}^s + m_r^s \cdot Z_r + \varepsilon_{r,j}^s \tag{2.9}$$

$$p_{r,j}^s = -\mathbf{u}_r^s \cdot \Delta \mathbf{r} + t_r + c \cdot \kappa_j \cdot d_{r,1} + \kappa_j \cdot I_{r,1}^s + m_r^s \cdot Z_r + e_{r,j}^s \tag{2.10}$$

$$B_{r,j}^s = N_{r,j}^s + b_{r,j} - b_j^s \qquad (2.11)$$

where $B_{r,j}^s$ is the real-valued undifferenced ambiguity.

At the epoch k, observational equations for all the satellites can be expressed as,

$$Y_k = A_k \cdot X_k + \varepsilon_{Y_k}, \ \varepsilon_Y \sim N(0, \sigma_Y^2) \qquad (2.12)$$

where the estimated parameters are,

$$X = (\Delta \mathbf{r}^T \ Z_r \ t_r \ d_{r,1} (I_{r,1}^s)^T (B_{r,1}^s)^T (B_{r,2}^s)^T)^T \qquad (2.13)$$

A sequential least squares filter is employed to estimate the unknown parameters for real-time processing. P_k denotes the weight matrix at epoch k,

$$\hat{X}_k = P_{\hat{X}_k}^{-1} \cdot (A_k^T P_k Y_k + P_{\hat{X}_{k-1}} \hat{X}_{k-1}) \qquad (2.14)$$

$$Q_{\hat{X}_k} = P_{\hat{X}_k}^{-1} = (A_k^T P_k A_k + P_{\hat{X}_{k-1}})^{-1} \qquad (2.15)$$

The increments of the receiver position $\Delta \mathbf{r}$ are estimated epoch by epoch without any constraints between epochs for retrieving rapid station movements. The tropospheric zenith wet delay Z_r is described as a random walk process with noise of about 2-5 $mm/\sqrt{hour}$. The receiver clock is estimated epoch-wise as white noise and the carrier-phase ambiguities $B_{r,j}^s$ are estimated as constant over time.

2.2.4 Real-Time Estimation of the Uncalibrated Phase Delay

From the carrier phase observational equation of (2.9), one can see that undifferenced phase ambiguities cannot be fixed to integers because of the existence of UPD. The real-valued ambiguities can be expressed by integer numbers plus the UPD at receiver and satellite as Eq. (2.11). Recent studies show, that the UPD can be estimated with high accuracy (Ge et al., 2008; Li and Zhang, 2012) or assimilated into clock parameters (Collins et al., 2008; Laurichesee et al., 2008).

In our procedure of real-time UPD estimation, the PPP float solution is firstly carried out for all stations of the reference network. In this PPP processing, the coordinates of the reference stations are fixed to well-known values, for example, determined from weekly solution in advance, so that the undifferenced (UD) float ambiguities could be obtained with higher quality.

Assume that we have a network with n stations (generally more than 15 stations for regional networks or more than 80 stations for a global network) tracking m satellites, the UD float ambiguities at each station are estimated as B_i, we have the observation equation in the form of Eq. (2.16) for these ambiguities (Li and Zhang, 2012),

$$\begin{bmatrix} B_1 \\ B_2 \\ \vdots \\ \vdots \\ B_n \end{bmatrix} = \begin{bmatrix} I & & & R_1 & S_1 \\ & I & & 0 & R_2 & S_2 \\ & & \vdots & & \\ & 0 & & I & & \\ & & & I & R_n & S_n \end{bmatrix} \begin{bmatrix} N_1 \\ N_2 \\ \\ N_n \\ b_r \\ b^s \end{bmatrix}, Q \quad (2.16)$$

where B_i is the UD float ambiguity vector for station i; N_i is the UD integer ambiguity vector for station i; b_r and b^s are the UPD for receivers and satellites; R_i and S_i are the coefficient matrices for receiver and satellite UPD respectively; Q is the covariance matrix of the UD float ambiguities; In matrix R_i one column with all elements is 1 and the others are zero. For matrix S_i each line is one element of −1, the others are zero.

Under the condition that all the integer ambiguities are exactly known and one UPD is fixed to zero, other UPD can be estimated by means of the least squares adjustment. Furthermore, for integer ambiguity resolution, the fractional part of UPD is sufficient instead of UPD itself as the integer part can be absorbed by the ambiguity parameter anyway. Therefore, we will not distinguish UPD and UPD fractional part hereafter.

Assume the receiver UPD at the first arbitrarily selected

station is zero, then the nearest integers of the UD ambiguities at this station are their integer ambiguities and the fractional parts are estimates of the related satellite UPD. Applying this satellite UPD to the common satellites of the next station, the corrected UD ambiguities should have very similar fractional part. The mean value of the fractional parts of all the common satellites is taken as UPD of the receiver. With this UPD, the UPD of the upcoming satellites, observed at the station, can be estimated. Repeating this procedure for all stations, we can have the approximate UPD for all receivers and satellites. After correcting the UD float ambiguities with the UPD, they should be very close to integers, thus ambiguity resolution can be attempted. Replacing integer ambiguity parameters with their fixed values in equation (2.16), the remaining parameters can be estimated and the UPD estimates are surely improved and will help to resolve more integer ambiguities.

The above described procedure can be done iteratively until no more integer ambiguities can be fixed. The UPD of the last iteration are the information needed for the user side for PPP ambiguity resolution. Finally, we broadcast the satellite UPD to PPP users together with orbit, clock and DCBs product streams, so that PPP with integer ambiguity resolution can be performed at a single station.

2.2.5 Ambiguity Resolution for Precise Point Positioning

Integer ambiguity resolution for PPP requires not only precise satellite orbit and high-rate satellite clock corrections but also the above-mentioned UPD. With the received UPD corrections, satellite UPD are removed at the user-end and the observational equation of (2.9) can be simplified as,

$$l_{r,j}^s = -\mathbf{u}_r^s \cdot \Delta \mathbf{r} + t_r + \lambda_j \cdot N_{r,j}^s + \lambda_j \cdot b_{r,j} - \kappa_j \cdot I_{r,1}^s + m_r^s \cdot Z_r + \varepsilon_{r,j}^s$$

(2.17)

The receiver UPD can be easily separated by adapting one

UD ambiguity to its nearest integer. Afterwards, the UD ambiguities have integer feature and the L1 and L2 ambiguities can be fixed simultaneously using integer estimation methods (see, e.g., Teunissen 1995; Xu et al., 2012). The ratio of the second minimum to the minimum quadratic form of residuals is applied to decide the correctness and confidence level of integer ambiguity candidate (the threshold for the ratio test is set to 3 as usual).

2.3 Accuracy Assessment in Real-Time Scenarios

For assessing the precision of our real-time product streams, fifteen real-time GPS stations (1Hz sampling), which are not used for server-side product generation, are selected as user stations. The PPP float solution and PPP with ambiguity resolution are carried out in parallel for these stations from day 325 to 334 2012 (November 20 to 29, 2012). The same orbit and clock corrections are used to feed the float PPP and the ambiguity-fixed PPP. Compared with GFZ final products (be regarded as truth), the 3D RMS (root mean squares) of orbit error is about 4.0 cm, the RMS of clock error is about 5.1cm, and the RMS of user range error (URE) is about 3.5cm. The station coordinates are estimated epoch-by-epoch without any constraints between epochs for testing the real-time kinematic performance. The estimated coordinates are compared with the coordinates derived from post-processed daily solutions to assess the positioning accuracy.

As a typical example, the differences of the estimated positions with respect to their daily solution for station A17D (Potsdam) on day 326 are exemplarily shown in Figure 2.3. Figure 2.3a shows the position differences (with respect to post-processed daily solution) of the PPP float solution in the north, east and up components. The differences are within 10 cm and 15 cm for the horizontal and vertical components, respectively. The vertical is, as expected, the noisiest component, due to the satellite constellation geometry and the high correlation between zenith tropospheric delay and the height component. There are long-term variations and short-term fluctuations in the position series even after a long convergence period of 24 hours (days 325). The position differences of the fixed solution are shown in Figure 2.3b. The improvement is significant, when compared with

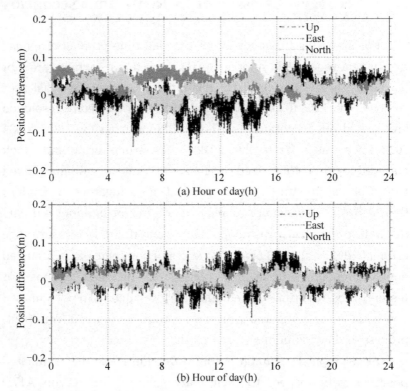

Figure 2.3 *Real-time kinematic PPP solutions for station A17D (Potsdam) on day 326 2012. (a) PPP without ambiguity-resolution; (b) PPP with fixed ambiguities. The north, east and up components are indicated by the green, red and black lines, respectively.*

the related float solution. In the ambiguity-fixed solution, the fluctuations are much smaller than those in the float solution and there are not obvious biases. A position accuracy of better than 5 cm and 10 cm in horizontal and vertical components is respectively achieved once the ambiguities are successfully fixed. Comparisons of the PPP float and fixed solutions for station BELL on days 326 are shown in Figure 2.4 and similar improvements are achieved in the ambiguity-fixed solution.

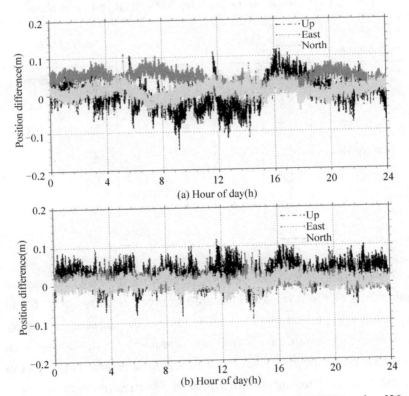

Figure 2.4   Real-time kinematic PPP solutions for station BELL on day 326 2012. (a) PPP without ambiguity-resolution; (b) PPP with fixed ambiguities. The north, east and up components are indicated by the green, red and black lines, respectively.

Chapter 2 High-Rate GNSS Seismology Using Real-Time PPP with Ambiguity Resolution

The RMS, standard deviation (STD) and mean bias of the position difference series are calculated as statistical indicators for the accuracy assessment. The statistical results of the position difference series of days 326 to 334 2012 (November 21 to 29, 2012) for stations A17D and BELL are summarized in Table 2.1. The RMS of the selected fifteen user stations for both PPP float and fixed solutions are also shown in Figure 2.5.

Table 2.1 Statistical results including RMS, STD and mean biases

Station & Accuracy	RMS (cm)		STD (cm)		Mean bias (cm)	
	Float	Fixed	Float	Fixed	Float	Fixed
North(A17D)	2.5	1.4	2.0	1.2	1.6	0.6
East(A17D)	3.7	1.3	1.9	1.0	3.2	0.8
Up(A17D)	5.7	2.8	4.7	2.5	-2.0	1.2
North(BELL)	1.8	1.4	1.5	1.2	1.0	0.5
East(BELL)	4.6	1.4	1.8	1.1	4.2	0.9
Up(BELL)	5.5	3.2	4.5	2.5	2.1	1.7

From Figure 2.5, the position RMS in north, east and vertical directions of the float solutions are improved from about 3, 5 and 6 cms to 1, 1 and 3 cms, respectively, by integer ambiguity resolution. Among these significant improvements the largest is in the east component. The STDs of float solutions are reduced from about 2 and 5 cms for the horizontal and vertical components to 1 and 2.5 cms, respectively. It is worth to mention that the biases are also decreased evidently by ambiguity resolution, especially in the east component from about 3-4 cms to several mm.

2.3 Accuracy Assessment in Real-Time Scenarios

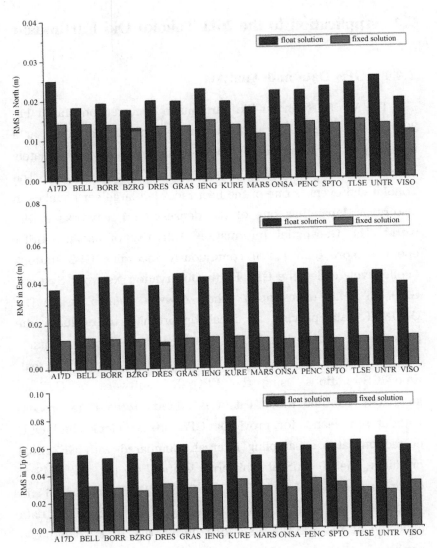

Figure 2.5 *RMS of the position differences of days 326 to 334 2012 (November 21 to 29, 2012) for the selected fifteen user stations (Global). The RMS of north, east and up components are shown in the top, middle and bottom sub-figure. The float solutions are in blue and the fixed ones in red.*

2.4 Application to the 2011 Tohoku-Oki Earthquake

2.4.1 GPS Data and Analysis

The Mw 9.0 Tohoku-Oki earthquake occurred on March 11, 2011 at 05:46:24 UTC in the north-western Pacific Ocean at a relatively shallow depth of 30 km, with its epicenter approximately 72 km east of the Oshika Peninsula of Tohoku, Japan. The Tohoku-Oki event is one of the best recorded large earthquakes in history as Japan has one of the densest GPS networks in the world. The Geospatial Information Authority of Japan (GSI) operates more than 1,200 continuously recording GPS stations (collectively called the GPS Earth Observation Network System, GEONET) all over Japan (http://www.gsi.go.jp/). The GEONET data provide an ideal opportunity to evaluate the performance of real-time PPP derived coseismic displacements.

We process 1 Hz data of about 80 globally distributed real-time IGS stations using the EPOS-RT software of GFZ in simulated real-time mode (data is edited in real-time process without pre-clean) for providing GPS orbits, clocks and UPD corrections at 5 s sampling interval. Compared with GFZ final products, the 3D RMS of orbit error is about 4.1 cm, the RMS of clock error is about 5.3 cm, and the RMS of URE is about 3.6 cm. Based on these corrections, we replayed the 1 Hz GPS data collected at the GEONET stations during the 2011 Tohoku-Oki earthquake. In the near future, it is difficult for most countries at threat from large earthquakes and tsunamis to afford such a dense GPS network as Japan's GEONET. To test the utility of a sparse GPS network for earthquake/tsunami early warning (Wright et al., 2012), sixty high-rate GPS stations are selected from the

GEONET for fault slip inversion in this study. The distribution of the selected GPS stations is shown in Figure 2.6.

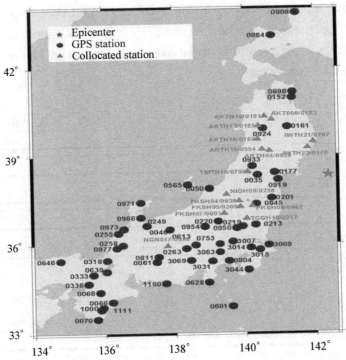

Figure 2.6 *Location of the 2011 Tohoku-Oki earthquake epicenter and the distribution of the selected high-rate GPS stations and strong motion stations. The epicenter is marked by the red star. The blue circles represent stations with GPS only, whereas the gray triangles are for stations with collocated GPS and strong motion seismometer.*

We calculated the RMS of the position differences of the sixty selected GPS stations over the 2 hours before the earthquake event for the PPP float and fixed solutions, respectively. It reveals that RMS in east, north and vertical of the float solutions of about 2.8, 4.4 and 9.7 cms are improved to 1.9, 1.9 and 4.0 cms,

27

correspondingly by integer ambiguity resolution. Similarly, we also calculated the RMS of the displacement-derived velocities over these two hours for the sixty GPS stations which are 2.7, 1.8 and 4.6 mms respectively in the north, east and up components. The displacement waveforms from PPP fixed solution and the velocities of five stations are shown as examples in Figures 2.7a and 2.7b, respectively.

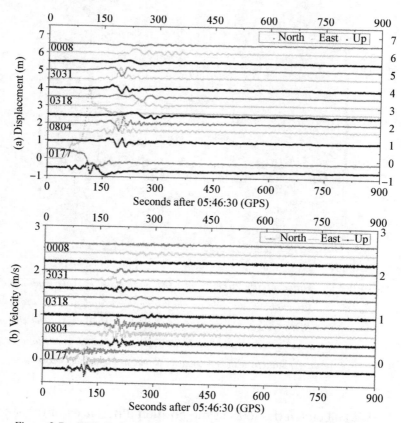

Figure 2.7 *PPP displacements and velocity waveforms at GPS stations 0008, 3031, 0318, 0804 and 0177 during the Tohoku-Oki earthquake on March 11, 2011. The north, east and up components are respectively shown by red, green and black lines.*

2.4.2 Comparing GPS and Seismic Waveforms

Japan has one of the densest seismometer networks in the world, and presently includes the F-Net, with 84 broadband stations; the K-NET, with 1,000 strong motion stations; the Hi-Net, with 777 high sensitivity stations (borehole installation); and the KiK-Net, co-located with the Hi-Net. We found about fifteen collocated GPS and strong motion station pairs (Figure 2.6). The strong motion recordings are firstly processed using the automatic empirical baseline correction scheme proposed by Wang et al. (2011). The velocity and displacement seismograms are then derived from the baseline-corrected strong motion recordings and compared with the GPS results at these collocated pairs.

The nearest station pair between the strong motion and GPS networks, that is, K-NET station AKT006 (40.2152° N, 140.7873° E) and the GEONET station 0183 (40.2154° N, 140.7873° E), being separated by 20 m. The PPP and seismic displacement waveforms from 0183 and AKT006 for the Tohoku-Oki earthquake are exemplarily compared in Figure 2.8. The 1 Hz ambiguity-fixed PPP displacements and 100 Hz seismic displacements are shown by red and black lines, respectively. The standard PPP float solution is also shown for comparisons with green lines.

The 900 s period of seismic shaking at 0183/AKT006 in the north, east and up components are respectively shown in the Figures 2.8a, 2.8b and 2.8c. Peak surface displacements at this station are about 1.0 m in the horizontal and 0.4 m in the vertical component. The comparisons of PPP and seismic displacements clearly show a high degree of resemblance, with aligned phase and very similar amplitudes of the dynamic component. The problem is that the permanent coseismic offsets in the PPP and

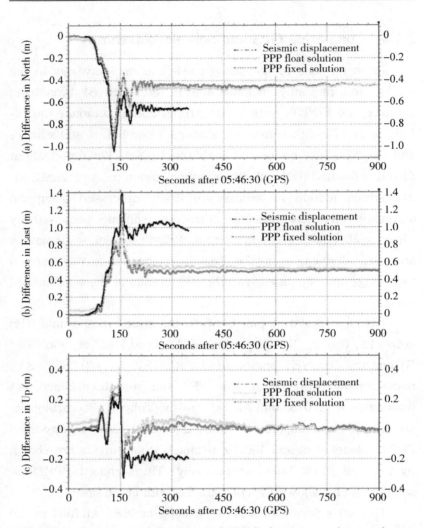

Figure 2.8 *Comparisons of the seismic and PPP displacement waveforms on the co-located AKT006 (seismic) and 0183 (GPS) stations during the 2011 Tohoku-Oki earthquake. The north, east and up components are respectively shown in the sub-figures a, b, and c. The real-time ambiguity-fixed PPP waveform, estimated from GPS observations, is shown by the red rectangle. The seismic displacement waveform is drawn by the black triangle. The PPP float solution is shown with the green cycle.*

seismic waveforms are very different. From the PPP waveforms, the permanent coseismic offsets of about 0.4 m, 0.5 m and few centimeters are respectively visible in Figures 2.8a, 2.8b and 2.8c. However, the corresponding coseismic offsets of seismic waveforms are about 0.6 m, 1.0 m and 0.2 m. Tilt and rotation of the seismic instrument lead to the baseline offsets in the seismic recordings, whereas the GPS receiver observes displacements directly and does not suffer from drift, clipping or instrument tilting. Although the empirical baseline correction has been applied to the seismic recordings, the accuracy of the permanent coseismic offsets derived from seismic waveforms is still not comparable with that of the GPS-derived coseismic offsets. The vertical GPS displacement is the noisiest component due to the satellite constellation geometry and the high correlation between zenith tropospheric delay and the height component (Wright et al., 2012).

The PPP and seismic displacement waveforms at GPS/seismic station pair 0986/NGN017 for the Tohoku-Oki earthquake are also shown in Figure 2.9. In general we found that the displacement waveforms, estimated from real-time ambiguity-fixed PPP and those provided by the accelerometer instrumentation are largely consistent.

The GPS velocity series is also compared with the velocity series integrated from the collocated accelerometers. The 1 Hz GPS velocity waveforms are derived from real-time ambiguity-fixed PPP. The 100 Hz seismic ones are obtained through single integration of accelerometer data. The velocity results at 0183/AKT006 and 0986/NGN017 are respectively shown in Figures 2.10 and 2.11 as examples where the velocity series derived from PPP fixed solution are shown by the red line and the corresponding seismic velocities are shown by the black line. From the Figures 2.10a, b and 2.11a, b, the velocity series in the horizontal

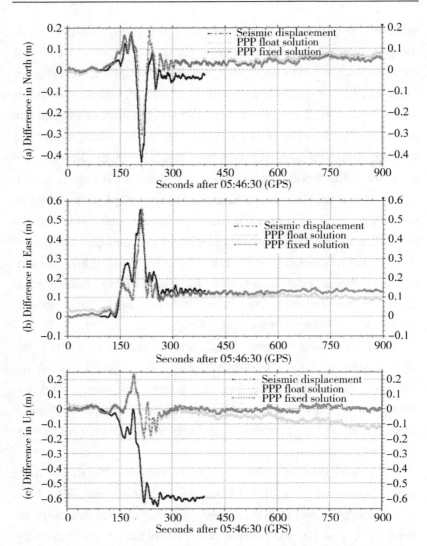

Figure 2.9 The seismic and PPP displacements on the co-located NGN017 (seismic) and 0986 (GPS) stations during the Tohoku-Oki earthquake on March 11, 2011. The north, east and up components are respectively shown in the sub-figures a, b, and c. The ambiguity-fixed PPP and seismic displacements are respectively shown by the red and black lines. The PPP float solution is drawn with the green line.

2.4 Application to the 2011 Tohoku-Oki Earthquake

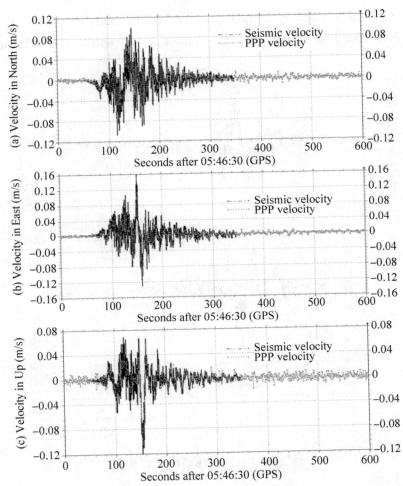

Figure 2.10 *Comparisons of the velocity series are derived from accelerometer and GPS on the co-located AKT006 (seismic) and 0183 (GPS) stations during the Tohoku-Oki earthquake on 11 March 2011. The north, east and up components are respectively shown in the sub-figures a, b, and c. The red line shows the velocity series derived from the PPP fixed solution, and the black line shows the corresponding velocity series integrated from accelerometer data.*

components show a high degree of consistency with the corresponding seismic results. Figures 2.10c and 2.11c indicate

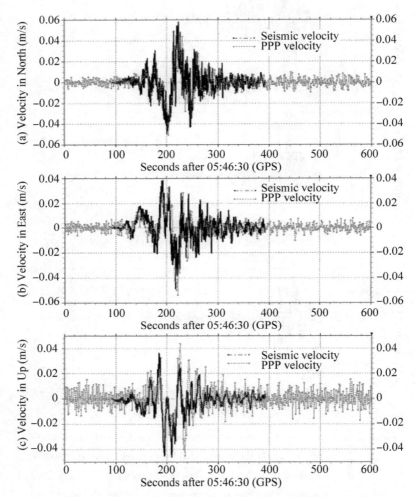

Figure 2.11 *The velocity series from the collocated 0986/NGN017 stations during the Tohoku-Oki earthquake on March 11, 2011. The north, east and up components are shown in the sub-figures a, b, and c respectively. The PPP velocity series are shown by the red line, while the seismic velocities are shown by the black line.*

that the vertical component is relatively noisy. A statistical analysis indicates that the RMS of the differences between PPP and seismic velocities are 2.5 cm, 2.2 cm and 3.2 cm in north, east, and vertical components, respectively.

2.4.3 Fault Slip Inversion

To further validate the ambiguity-fixed PPP, we apply the PPP displacements to fault slip inversion of the Tohoku-Oki earthquake. The inversions are carried out using the code SDM written by Dr. R. Wang based on the constrained least-squares method, which has been used in a number of recent publications for analyzing GPS, InSAR and strong motion based co-and post-seismic deformation data (Diao et al., 2010; Wang et al., 2013; Li et al., 2013c). For simplicity of numerical analysis, the fault plane is represented by a number of small rectangular dislocation patches with uniform slip. The observed displacement data are related to the discrete fault slips through Green's functions of the earth model, which are calculated using linear elastic dislocation theory. For the discrete slips to be an adequate representation of the true continuous slip distribution, the patch size must be reasonably small. In fact, if the available data do not include enough information for determining the slip distribution with the desired resolution, the inversion system becomes underdetermined. To overcome the problem of nonuniqueness and instability inherent in such an underdetermined system, priori conditions (fixed fault geometry and restricted variation range for the rake angle) and physical constraints (smooth spatial distribution of slip or stress drop) are considered.

We derive the spatial distributions of fault slip using the coseismic displacements obtained from the real-time PPP float solution, real-time PPP fixed solution, and the post-processed

ARIA solution, respectively. The post-processed ARIA solution provided by the ARIA team at JPL and Caltech (Simons et al., 2011) is depicted as reference. In the same way as done by Wang et al. (2013), we employ a slightly curved fault plane, parallel to the assumed subduction slab. The dip angle increases linearly from 10° on the top (ocean bottom) to 20° at about 80 km depth. To avoid any artificial bounding effect, a large enough potential rupture area of 650 km ×300 km is used. The upper edge of the fault is located along the trench east of Japan, on the boundary between the Pacific plate and the North America plate. The patch size is 10 km × 10 km. The rake angle determining the slip direction at each fault patch is allowed to vary between 90°± 20°. Green's functions are calculated based on the CRUST2.0 model (Bassin et al., 2000) in the concerning area. In the inversion, the data is weighted twice as much for the two horizontal components as for the vertical component.

The inverted fault slip distributions are shown in Figure 2.12, and the comparisons of observed and synthetic displacements on horizontal and vertical components are shown in Figure 2.13. The three inversions result in scalar seismic moments of 2.1×10^{22}, 3.4×10^{22}, 3.4×10^{22} Nm and they are equivalent to moment magnitudes of Mw 8.82, Mw 8.96, and Mw 8.96, respectively. The maximum slip of the three inversion results are 16.7 m, 21.2 m and 21.1 m, respectively. The PPP float solution leads to underestimations of about 38% on the scalar seismic moment and about 21% on maximum fault slip, which may lead to an underestimation on tsunami scale. PPP fixed solution is quite consistent with post-processed ARIA solution, and is much better than PPP float solution. The PPP float solution shows a circular rupture without obvious rupture propagation and direction, while the PPP fixed solution and post-processed ARIA solution show

2.4 Application to the 2011 Tohoku-Oki Earthquake

that the rupture mainly propagates along the fault up-dip direction and toward the sea bed. Integer ambiguity resolution brings the PPP moment magnitude into agreement with the post-processed value. The moment magnitude of the earthquake we estimated (Mw 8.96) in this study is similar to the moment solutions of about Mw 9.0, estimated by the USGS, and slightly smaller than Mw 9.1 of Global CMT. The inversion results of real-time PPP fixed solution and the post-processed ARIA solution are quite similar to each other not only in the moment magnitude, but also in the slip distribution pattern. Overall, the comparison of the three inversion results shows that integer ambiguity resolution in PPP is beneficial for fault slip inversion and the moment magnitude estimation. The PPP fixed solution can provide a reliable estimation of earthquake magnitude and even of the fault slip distribution in real time.

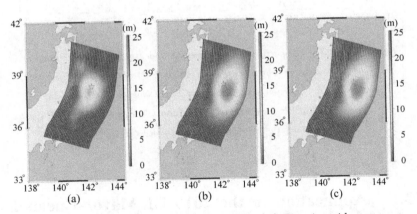

Figure 2.12 *The inverted fault slip distributions. (a) Inversion with permanent displacements obtained from real-time PPP float solution; (b) Inversion with real-time PPP fixed solution; (c) Inversion with post-processed ARIA solution.*

Chapter 2 High-Rate GNSS Seismology Using Real-Time PPP with Ambiguity Resolution

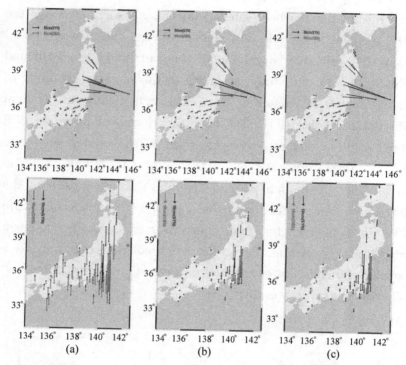

Figure 2.13 *The comparisons of the observed and synthetic displacements on horizontal components, and on vertical components, respectively. a) Inversion with permanent displacements obtained from real-time PPP float solution; (b) Inversion with real-time PPP fixed solution; (c) Inversion with post-processed ARIA solution.*

2.5 Application to the 2010 El Mayor-Cucapah Earthquake

2.5.1 GPS Data and Analysis

The 2010 Mw 7.2 El Mayor-Cucapah earthquake (April 4,

2.5 Application to the 2010 El Mayor-Cucapah Earthquake

2010, 22:40:42 UTC), struck Baja California approximately 65 km south of the US-Mexico border. This earthquake ruptured along the principal plate boundary between the North American and Pacific plates with a shallow focal depth of about 10 km. Surface rupture of this earthquake extended for about 120 km from the northern tip of the Gulf of California northwestward nearly to the international border, with breakage on a series of faults occupying a general NW-SE zone. It caused significant ground motions at distances up to several hundred kilometers from the epicenter.

Most of the broadband seismometers close to the epicenter clipped in this event, strong motion seismometers and high-rate GPS receivers are the two major candidate instruments to detect the surface displacement. The UNAVCO Plate Boundary Observatory (UNAVCO-PBO) of EarthScope is a geodetic observatory designed to characterize the three-dimensional strain field across the active boundary zone between the Pacific Plate and the western United States. The El Mayor-Cucapah Earthquake was well recorded not only by strong motion stations but also by high-rate GPS receivers with a 5 Hz sampling rate at the PBO stations. This event is one of the best examples in California of a large earthquake for which abundant high-rate GPS and strong motion records are available (Allen and Ziv, 2011). With the real-time corrections generated by the EPOS-RT software (the RMS of orbit error, clock error, and URE are about 4.3 cm, 5.6 cm, and 3.7 cm, respectively), we replay the 5 Hz GPS data collected by 30 nearfield UNAVCO-PBO stations during the El Mayor-Cucapah earthquake. These stations are used for fault slip inversion and their distribution is shown in Figure 2.14.

Chapter 2 High-Rate GNSS Seismology Using Real-Time PPP with Ambiguity Resolution

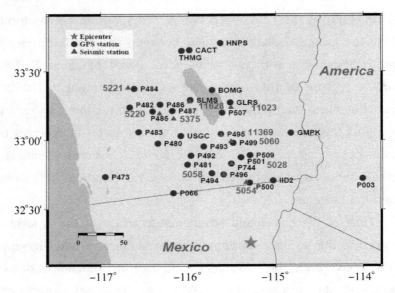

Figure 2. 14 *Location of the El Mayor-Cucapah earthquake and the distribution of the selected high-rate GPS and strong motion stations. The epicenter of the El Mayor-Cucapah earthquake is marked by the red star. The blue circles represent the GPS stations, and the gray triangles are strong motion stations.*

We calculated the RMS values of 2 hours (before the earthquake event) position series (after convergence) of the 30 GPS stations. The RMS values of PPP float solution are found to be 2.2 cm, 4.1 cm, and 7.6 cm respectively in north, east and up components. PPP ambiguity resolution can improve the accuracy to 1.8 cm, 1.9 cm and 3.9 cm in the corresponding components. We calculated the RMS values of two hours (before the earthquake event) velocity series for the 30 GPS stations. The RMS values are found to be 1.2 cm, 0.7 cm and 4.0 cm respectively in the north, east and up components.

2.5.2 Comparing GPS and Seismic Waveforms

Some of the GPS stations are co-located with seismic stations from the Southern California Seismic Network (SCSN) operated by the USGS (U.S. Geological Survey) and Caltech (Figure 2.14). Almost all broadband velocity network instruments in California also have an accelerometer in order to record large magnitude events when the velocity instruments are likely to clip. The accelerometers do not clip, and velocity and displacement waveforms can be obtained through single and double integrations, respectively. The velocity and displacement seismograms in this study are provided by California Geological Survey (CGS/CSMIP, http://strongmotioncenter.org/). The baseline offsets are removed by applying a high-pass filter at the price of low-frequency information loss, including the loss of permanent station offsets.

The PPP and seismic displacement waveforms from P744 and 5028 for the El Mayor-Cucapah earthquake are exemplarily compared in Figure 2.15. The 5 Hz ambiguity-fixed PPP displacements and 200 Hz seismic displacements are shown by the red and black lines respectively. The standard PPP float solution is also shown for comparisons with the green line. In the Figure 2.15a, we show the entire period of seismic shaking at P744/5028 in the north component. The north component shows very similar amplitudes of the dynamic component. From the PPP waveforms, the permanent coseismic offsets of about 0.1 m are respectively visible in Figure 2.15a, while permanent coseismic offsets are lost in the seismic waveforms. The Figure 2.15b displays an excellent agreement of seismic and ambiguity-fixed PPP displacements in the east component within few centimeters. From the Figure 2.15c, the vertical component has been evidently improved by PPP ambiguity resolution, but it is still the noisiest component.

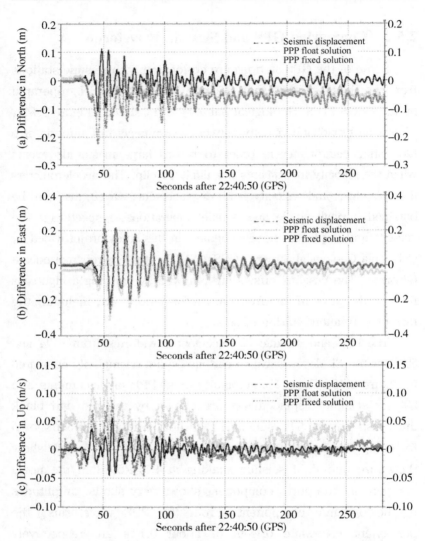

Figure 2.15 *The seismic and PPP displacements on the co-located 5028 (seismic) and P744 (GPS) stations during the El Mayor-Cucapah earthquake on April 4, 2010. The north, east and up components are respectively shown in the sub-figures a, b, and c. The ambiguity-fixed PPP and seismic displacements are respectively shown by the red and black lines. The standard PPP float solution is also drawn for comparisons with the green line.*

The GPS velocity series are also compared with those, integrated from the collocated accelerometers. The 5 Hz GPS velocity waveforms are derived from real-time ambiguity-fixed PPP. The 200 Hz seismic ones are obtained through single integration of accelerometer data. The velocity results at P744/ 5028 are shown in Figure 2.16. The velocity series derived from PPP fixed solution are shown by the red line, while the corresponding seismic velocities are shown by the black line. From the Figures 2.16a and 2.16b, the velocity series in the horizontal components show a high degree of consistency with the corresponding seismic results. The Figure 2.16c indicates that the vertical component is relatively noisy. A statistical analysis indicates that the RMS of the differences between PPP and seismic velocities are 2.1 cm, 2.0 cm and 3.1 cm in the north, east, and vertical components, respectively.

2.5.3 Fault Slip Inversion

We derive the spatial distributions of fault slip using the permanent coseismic displacements obtained from the real-time PPP float solution, real-time PPP fixed solution, and the post-processed daily solution (the difference between the day before the earthquake and the day after the earthquake), respectively. The fault geometric parameters (strike 313°/dip 88°) are adopted from the global centroid moment tensor (GCMT) solution of the earthquake. The rake angle (slip direction relative to the strike) is allowed to vary ±20° around the GCMT solution of 186°. The fault is given to be 130 km along the strike and 20 km wide down the dip, which is then divided into $26 \times 4 = 104$ sub-faults. To avoid unreasonable slip patterns, the smoothing constraint is imposed. In the inversion, the data is weighted twice as much for the two horizontal components as for the vertical component.

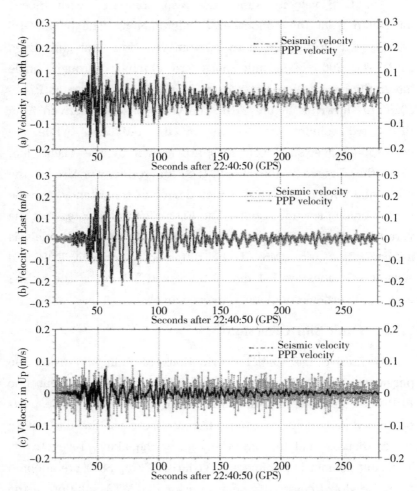

Figure 2.16 *The velocity series on the collocated P744/5028 stations during the El Mayor-Cucapah earthquake on April 4, 2010. The north, east and up components are shown in the sub-figures a, b and c respectively. The PPP velocity series are shown by the red line, while the seismic velocities are shown by the black line.*

The inverted fault slip distributions are shown in Figure 2.17, and the comparisons of observed and synthetic displacements on

2.5 Application to the 2010 El Mayor-Cucapah Earthquake

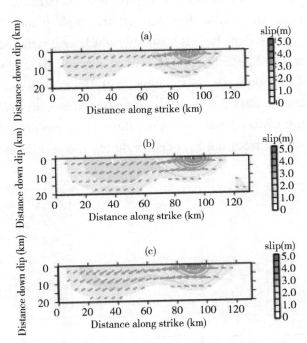

Figure 2.17 *The inverted fault slip distributions. (a) Inversion with permanent displacements obtained from real-time PPP float solution; (b) Inversion with real-time PPP fixed solution; (c) Inversion with post-processed daily solution.*

horizontal and vertical components are shown in Figure 2.18. The three inversions result in scalar seismic moments of 5.4×1019, 7.6 ×1019 and 7.6×1019 Nm, equivalent to moment magnitudes of Mw 7.09, 7.19, and 7.19, respectively. The PPP float solution leads to an underestimation of about 30% of the scalar seismic moment, and PPP fixed solution significantly improves it. The inversion results of real-time PPP fixed solution and post-processed daily solutions are quite similar with each other not only in the moment magnitude, but also in the slip distribution pattern. The major slip

45

area occurred at a very shallow depth (near the surface) at about 90 km along the strike direction on the fault plane. The rake variation shows that there is a purely right lateral strike slip at the northwest of the fault, and a minor normal fault component occurs at the south east of the fault. With the consideration of the hypocentral location, we can confirm that this earthquake is an asymmetric bilateral rupture event: the rupture mainly propagates northwestward from the hypocenter during the source process.

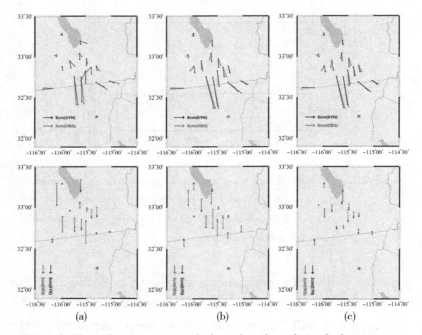

Figure 2.18　*The comparisons of observed and synthetic displacements on horizontal components, and on vertical components, respectively. a) Inversion with permanent displacements obtained from real-time PPP float solution; (b) Inversion with real-time PPP fixed solution; (c) Inversion with post-processed daily solution.*

2.6 Conclusions

Based on the IGS real-time infrastructure, several IGS RTACs demonstrated the capacity of the real-time global PPP services by providing precise orbit and clock products under the IGS RTPP framework. The standard PPP service is recently significantly improved in both reconvergence time and positioning accuracy by applying integer ambiguity resolution and using raw observation with ionosphere delays as parameters with proper constraints. These recent improvements are implemented into the real-time PPP service, established by GFZ as one of the IGS RTAC with integer ambiguity resolution for precise positioning and geophysical applications.

In order to assess the precision of the real-time PPP service, fifteen real-time GPS stations which are not used for product generation are selected as user stations. From the experimental results, the long-term variations and short-term fluctuations in the float PPP solutions are significantly reduced by applying integer ambiguity resolution, especially in east component. On average, the position RMS of the float solutions in east, north and up directions are reduced from about 3, 5 and 6 cms to that of the fixed solution of about 1, 1 and 3 cms, respectively. This improvement is also demonstrated by the comparison of the positions derived from float and fixed solutions over the two hours period before the 2011 Tohoku-Oki and El Mayor-Cucapah earthquakes.

The 2011 Mw 9.0 Tohoku-Oki earthquake (March 11, 2011, 05:46:23 UTC) in Japan and the 2010 Mw 7.2 El Mayor-Cucapah earthquake (April 4, 2010, 22:40:42 UTC) in northern Baja California provide us with real events to evaluate the performance of real-time PPP derived coseismic displacements. Both of these two events were well recorded not only by strong motion stations but also by high-rate GPS receivers. PPP displacements are compared with the seismic displacements at the collocated stations. The comparison shows a high degree of resemblance between PPP and seismic displacements, with aligned phase and very similar amplitudes of the dynamic component. The difference is that permanent coseismic offsets are clearly visible in the PPP waveforms, but accurate permanent coseismic offsets are not available in the seismic waveforms because of the baseline offsets caused by tilt and rotation of the seismic instrument. In general the displacement waveforms, estimated from real-time ambiguity-fixed PPP and those provided by the accelerometer instrumentation are largely consistent in the dynamic component within few centimeters. The results also indicate that integer ambiguity resolution improves the accuracy of real-time PPP displacements significantly.

We derive the spatial distributions of fault slip for the 2011 Tohoku-Oki earthquake and the 2010 El Mayor-Cucapah earthquake using the coseismic displacements obtained from the real-time PPP float solution, real-time PPP fixed solution, and post-processed solution, respectively. In the case of the 2010 El Mayor-Cucapah earthquake, the three inversions result in scalar seismic moments of 5.4×10^{19} Nm, 7.6×10^{19} Nm, 7.6×10^{19} Nm, equivalent to moment magnitudes of Mw 7.09, Mw 7.19, and Mw 7.19, respectively. The PPP float solution leads to an underestimation of about 30% of the scalar seismic moment, and PPP fixed solution significantly improves it. For the 2011 Tohoku-Oki earthquake, the three inversions result in scalar seismic moments of 2.1×10^{22} Nm, 3.4×10^{22} Nm, 3.4×10^{22} Nm, equivalent to moment magnitudes of Mw 8.82, Mw 8.96, and Mw 8.96, respectively. The PPP float solution leads to an underestimation of about 38% of the scalar seismic moment, integer ambiguity resolution brings the PPP moment magnitude into agreement with the post-processed value. The PPP fixed solution can provide a reliable estimation of earthquake magnitude and even of the fault slip distribution in real time and become complementary to existing seismic EEW methodologies. The real-time ambiguity-fixed PPP module can be embedded into high-rate GPS receiver firmware and be incorporated into EEW systems especially for regions at threat from large magnitude earthquakes and tsunamis.

Chapter 3   Augmented PPP for Seismological Applications Using Dense GNSS Networks

3.1 Introduction

Appropriate and precise GPS real-time data analysis is crucial for the use of the network data for hazard monitoring. Currently, the relative baseline/network positioning technique is predominantly used for this purpose. For moderate-to-short baselines, integer ambiguity resolution can be achieved within a few seconds and sometimes with only one observational epoch to achieve a high positioning accuracy of a few cm. For the relative positioning technique, GPS data from a network is analyzed simultaneously to estimate station positions. This is a significant limitation for the challenging simultaneous and precise real-time analysis of GPS data from hundreds or thousands of ground stations. Furthermore, intermittent station dropouts complicate the network-based relative positioning. Relative positioning also requires a local reference station, which might itself be displaced during a large seismic event, resulting in misleading GPS analysis results. The reference station should be sufficiently far from the focal region, but must also be part of a sub-network that has relatively short baselines.

Alternatively, precise point positioning (PPP) can provide "absolute" displacements with respect to a global reference frame (defined by the satellite orbits and clocks) using a single GPS receiver. It is more flexible than the relative positioning technique and is widely used for hazard monitoring. However, the PPP method requires a long convergence period of about 20 minutes after receiver activation or after serious and/or long signal interruption for most of the GPS satellites. The worst case scenario for the GPS component of an earthquake/tsunami monitoring system would be a power failure during the disaster, which would reduce the usefulness of the PPP based displacement solution because of the time required for reconvergence. To avoid this major disadvantage, the PPP regional augmentation (Li et al., 2011) has been developed by making use of atmospheric corrections from a regional reference network to achieve nearly instantaneous ambiguity resolution. But the regional monitoring stations could also be displaced by the earthquake. Therefore the current PPP regional augmentation, in which the reference stations are assumed being in static mode and even with known coordinates for generating atmospheric corrections (Li et al., 2011) or pre-fit undifferenced observation residuals (Ge et al., 2012), could not be used for earthquake monitoring.

This is our motivation to propose here a novel method for fast or even instantaneous positioning, making full use of the currently available global PPP service and regional GPS monitoring networks. We estimate coordinates of all monitoring stations in kinematic mode to avoid the effects of the earthquake induced-displacements on atmospheric corrections. The derived atmospheric corrections at the stations with fixed ambiguities then can be provided to other monitoring stations for instantaneous ambiguity resolution, so that precise displacements can always be achieved within a few seconds. The series of displacements, derived using the proposed method, will be uninterrupted even in case of a break in tracking (loss of signal lock, cycle slips, or data gaps) due to a power outage or similar disruption. This is a considerable advantage for hazard monitoring application. The new method does not depend on a specific reference station and therefore the analysis results will not be affected by simultaneous shaking of any particular station. It also has better flexibility and efficiency compared to complicated network/subnetwork analysis. We demonstrate the advantages of the novel augmented PPP approach using 1 Hz GEONET data, collected during the Tohoku-Oki earthquake (Mw 9.0, 11 March, 2011) in Japan.

3.2 Augmented PPP Approach

Successful resolution of integer-cycle carrier-phase ambiguities is a prerequisite to achieve the most precise position estimates with GPS by transforming precise but ambiguous phase range measurements into precise unambiguous measurements (Blewitt, 1989; Dong and Bock, 1989). For relative positioning, the uncalibrated phase delays (UPD) are removed by the application of the double-difference (DD) technique and thus the phase ambiguity can be fixed to integers (Dong and Bock, 1989). The atmospheric delays are also mostly eliminated in case of moderate-to-short baselines, so that integer-cycle phase ambiguities can be fixed within few seconds. Recent studies show that the UPDs can be estimated with high accuracy and reliability from a global reference network and transferred to the GPS monitoring station to allow resolution of the ambiguities without differencing (Li et al., 2013; Ge et al., 2008). Several international GNSS service (IGS) analysis centers provide GPS orbit, clock, and UPD data products to allow real-time PPP use enabling ambiguity resolution anywhere in the world (Dow et al., 2009; Ge et al., 2012; Loyer et al., 2012). However, PPP still needs a comparatively long (re)convergence time of approximate 20 minutes to achieve reliable integer ambiguity resolution because precise atmospheric delay models cannot be derived from such a sparse global reference network (Li et al., 2013).

An increasing number of regional GPS monitoring networks are installed around the world for precise navigation and geophysical applications, especially in seismically active regions (e.g. Japan, Western North America, Greece, and Chile). One possible solution for achieving fast ambiguity resolution in PPP is to retrieve the atmospheric delays as corrections from data of these dense regional networks. By applying the UPD corrections, the integer un-differenced ambiguities on the L1 and L2 frequencies can be fixed in PPP mode at all regional monitoring stations. The atmospheric corrections of the ionospheric slant and tropospheric zenith wet delay then can be derived from the PPP fixed solution as

$$l_{r,j}^s - m_r^s \cdot Z_r + u_r^s \cdot \Delta r = -l_{r,j}^s - t^s + t_r + \lambda_j(b_{r,j} - b_j^s) + \lambda_j N_{r,j}^s + \varepsilon_{r,j}^s \tag{3.1}$$

$$-I_{r,j}^s - m_r^s \cdot Z_r + u_r^s \cdot \Delta r = -p_{r,j}^s - t^s + t_r + c(d_{r,j} + d_j^s) + e_{r,j}^s \tag{3.2}$$

Where, $l_{r,j}^s$, $p_{r,j}^s$ denote phase and code observation minus computation (OMC) from satellite s to receiver r at frequency j; u_r^s is the unit direction vector from site to satellite; Δr denotes the increments of the receiver positions; Z_r denotes the tropospheric zenith wet delay; m_r^s is the wet mapping function; t^s and t_r are the clock errors; λ_j is the wavelength; $b_{r,j}$ is the receiver-dependent uncalibrated phase delay; b_j^s is the satellite-dependent UPD; $d_{r,j}$

and d_j^s are the code biases; $I_{r,j}^s$ is the ionospheric delay; $N_{r,j}^s$ is the integer phase ambiguity; $e_{r,j}^s$ and $\varepsilon_{r,j}^s$ are the measurement noise terms of the pseudo-range and carrier phase.

This procedure is very flexible and computational efficient to be applied even for monitoring networks with a large number of stations as the atmospheric corrections are derived for each station individually. Because regional monitoring stations themselves could be displaced by the earthquake, the coordinates are estimated in kinematic mode to avoid the effects of earthquake induced-displacements on the atmospheric corrections that are generated. The constraints imposed on the kinematic coordinates of adjacent epochs are fine-tuned by using an adaptive filter (Yang et al., 2001) in real-time to strengthen the solution. Usually atmospheric delay is rather stable over short periods and can be represented by a constant or a linear function. Therefore, even in periods of strong shaking, station position and atmosphere are distinguishable in parameter estimation because of the significant difference in their temporal characters.

A polynomial model can be used to represent the derived atmospheric corrections on small regional scales. Here three or more nearby monitoring stations are selected as augmenting stations for each monitoring station, and the atmospheric corrections of the selected augmenting stations are interpolated by a Linear Combination Method (LCM) (Han, 1997) as

$$\sum_{i=1}^{n} \alpha_i = 1, \sum_{i=1}^{n} \alpha_i (\hat{X}_m - \hat{X}_i) = 0, \sum_{i=1}^{n} \alpha_i^2 = Min \quad (3.3)$$

$$\hat{v}_m = \sum_{i=1}^{n} \alpha_i \hat{v}_i \quad (3.4)$$

Where, n is the number of selected augmenting stations; m and i are indices for the monitoring and the selected augmenting stations, respectively; α_i denotes the interpolation coefficient; $\hat{X}_m$

and $\hat{X}_i$ are the station coordinates in the local horizontal plane system; ΔX_{im} and ΔY_{im} are the plane coordinate differences between the monitoring and augmenting station; $\hat{v}_i$ is the ionospheric or tropospheric delay; $\hat{v}_m$ is the interpolated ionospheric or tropospheric delay at the monitoring station.

For regional reference networks with moderate-to-short baselines (few tens of kilometers inter-station distance) cm-level accuracy can be achieved for the interpolated atmospheric delay corrections. The precise interpolated atmospheric corrections are imposed as a strong constraint on the related parameters of the monitoring station, while the coordinates are estimated in kinematic mode. Assuming that r_1 to r_n are selected as augmenting stations for interpolating corrections for the monitoring station r_m. The ionospheric slant delay parameter for an individual satellite s_i is constrained to the interpolated correction as

$$I_{r_m}^{s_i} - \tilde{I}_{r_1, r_2 \cdots r_n}^{s_i} = w_I, \quad w_I \sim N(0, \sigma_{w_I}^2) \quad (3.5)$$

And the constraint for the zenith wet delay parameter is

$$Z_{r_m} - \tilde{Z}_{r_1, r_2 \cdots r_n} = w_T, \quad w_T \sim N(0, \sigma_{w_T}^2) \quad (3.6)$$

Where $I_{r_m}^{s_i}$ denotes the slant ionospheric delay from station r_m to satellite s_i; $\tilde{I}_{r_1, r_2 \cdots r_n}^{s_i}$ is the interpolated ionospheric correction; Z_{r_m} denotes the zenith wet delay for station r_m, and $\tilde{Z}_{r_1, r_2 \cdots r_n}$ is the interpolated correction. w_I and w_T are the biases between the true and the interpolated atmospheric corrections. The statistical processes of w_I and w_T are zero mean white processes with variance of $\sigma_{w_I}^2$ and $\sigma_{w_T}^2$ for the ionospheric and tropospheric delays, respectively.

By adding this precise atmospheric delay model to the orbit, clock and UPD products used in global PPP ambiguity resolution, instantaneous ambiguity resolution is achievable at the monitoring station, so that the augmented PPP can have ambiguity resolution performance equivalent to relative positioning. It should be mentioned that the selection of augmenting stations is critical, as atmospheric corrections can only be derived from augmenting stations at which the ambiguity resolution is successfully achieved.

3.3 Application of Augmented PPP Approach and Results

The 2011 Mw 9.0 Tohoku-Oki earthquake (11 March 2011, 05:46:23 UTC) in Japan is one of the best GPS recorded large earthquakes, as Japan has one of the densest GPS networks in the world. The Geospatial Information Authority of Japan (GSI) operates more than 1,200 continuously observing GPS stations (collectively called the GPS Earth Observation Network System) all over Japan. The geographical distribution of the stations is indicated in Figure 3.1. The use of the GEONET data provides an excellent opportunity to evaluate the performance of our novel PPP analysis method. We replayed all the 1 Hz GPS data collected by the GEONET stations during the 2011 Tohoku-Oki earthquake using the augmented PPP method in simulated real-time mode.

First we process 1 Hz GPS ground tracking data of about 80-90 globally distributed real-time IGS stations using the GFZ's EPOS-RT software in simulated real-time mode for providing GPS orbits, clocks and uncalibrated phase delays (UPD) (Li et al., 2013) corrections at a 5 s sampling interval. Using the orbits, clocks and UPD data, the integer ambiguities are fixed in PPP mode for all of the GEONET stations and atmospheric corrections are derived on an individual station basis. For each GEONET station, three nearby GPS stations are selected as augmenting stations. In addition to the orbit, clock and UPD data products

3.3 Application of Augmented PPP Approach and Results

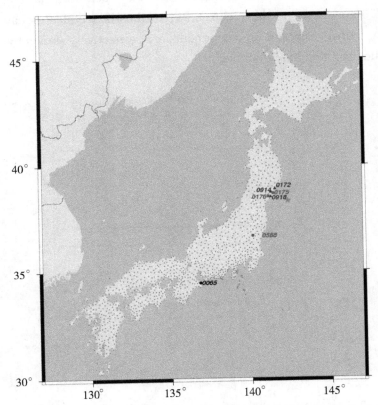

Figure 3.1　*Location of the 2011 Tohoku-Oki earthquake epicenter and the distribution of the high-rate GPS sites. The epicenter is marked by the red star. The brown circles represent GPS sites. The black diamond represents the reference site of relative positioning analysis. The purple rectangles represent the sites of the time series examples. This figure is drawn using GMT software.*

from the global PPP service, the atmospheric corrections of the augmenting stations are interpolated and imposed as a constraint on related parameters. Then instantaneous ambiguity fixing is performed independently at each epoch. The displacement

Chapter 3   Augmented PPP for Seismological Applications Using Dense GNSS Networks

waveforms, derived from augmented PPP solution, at station 0176 are shown in the Figure 3.2a, to illustrate typical behavior. The north, east and up components are respectively shown by the blue, red and black curves.

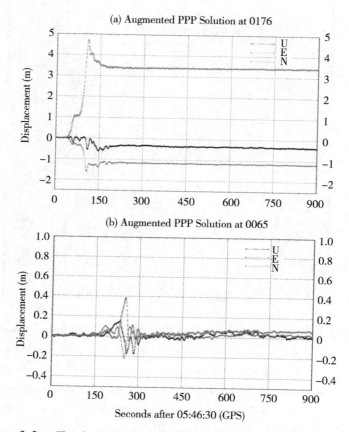

Figure 3.2   *The displacement waveforms derived from the augmented PPP solution. The north, east and up components are shown by the blue, red and black curves, respectively. a) The displacement waveforms at station 0176; b) The displacement waveforms at the reference station 0065.*

3.3 Application of Augmented PPP Approach and Results

It is found that there are significant GPS data gaps or cycle slips during the seismic shaking at some GEONET sites (e.g., 0175, 0588, etc.). There is for example a data gap of about 2 min at station 0175, which starts at epoch 05:47:21 and ends 05:49:27 (GPS time, GPST). The displacements from the augmented PPP solution for station 0175 are shown in Figure 3.3a. Stations 0172, 0914 and 0918 are selected as the augmenting stations for 0175. The estimated ionospheric corrections during seismic shaking at the augmenting stations are illustrated in Figure 3.4a and Figure 3.5a. The estimated zenith wet delays during the 600 s seismic shaking period are respectively 5.4 ± 0.1 cm, 6.5 ± 0.1 cm and 6.4 ± 0.1 cm at three augmenting stations. With the atmospheric corrections, retrieved from the augmenting stations, the atmospheric delays for 0175 are interpolated using the linear combination method. The resulting interpolations are compared with the estimated values at 0175 in order to assess the accuracy of the interpolation. Figure 3.4b and Figure 3.5b show the ionospheric differences between interpolated and estimated values. The differences are found to be smaller than 5 cm. The tropospheric interpolation error is about 0.26 cm. We found that the interpolated atmospheric corrections are accurate enough for rapid ambiguity resolution.

We also derive displacement waveforms of all GEONET stations from the relative positioning (RP) and global PPP solutions, and compare them with the augmented PPP solution. The global PPP displacements for station 0175 are shown in Figure 3.3b. In the global PPP solution, the displacement series shows a large disturbance after the data gap that is caused by the convergence sequence for fixing the PPP ambiguities (about 20 mins). This unstable behavior is an unavoidable problem for a real-time PPP use as the sparse global reference network

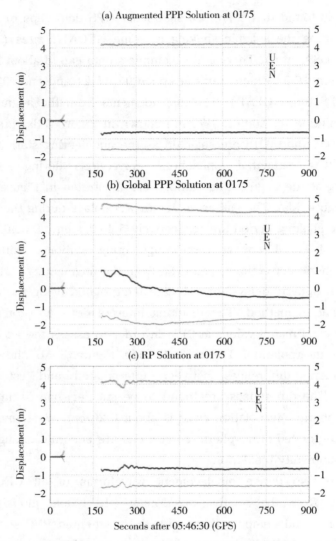

Figure 3.3 *Comparisons of the displacement waveforms derived from augmented PPP, global PPP and relative positioning solutions. The north, east and up components are shown by the blue, red and black curves, respectively. a) The displacements derived from augmented PPP solution at station 0175, which has a data gap of about 2 minutes during the seismic shaking; b) The displacements derived from the global PPP solution at station 0175; c) The displacements derived from relative positioning at station 0175, with 0065 as reference station.*

3.3 Application of Augmented PPP Approach and Results

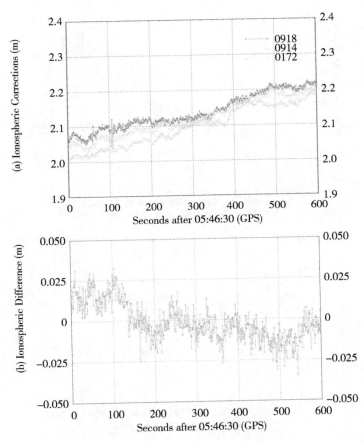

Figure 3. 4 *Ionospheric corrections at augmenting stations and ionospheric interpolation errors. a) The estimated ionospheric corrections for GPS satellite PRN 15 at the augmenting stations 0172, 0914 and 0918 during seismic shaking; b) Ionospheric interpolation errors of PRN 15 for the station 0175.*

employed cannot provide accurate atmosphere delays for fast ambiguity resolution. The relative positioning solution for the station 0175 is also shown in Figure 3. 3c. For the relative positioning analysis, we adopt the same reference station 0065 as

Chapter 3 Augmented PPP for Seismological Applications Using Dense GNSS Networks

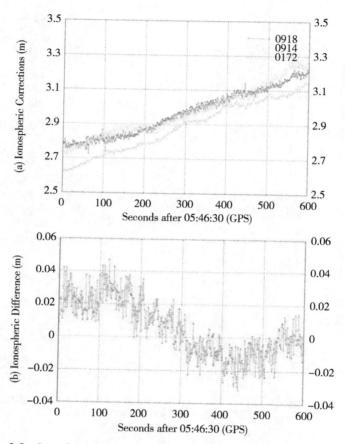

Figure 3.5 *Ionospheric corrections for the augmenting stations and the ionospheric interpolation errors. a) The estimated ionospheric corrections for GPS satellite PRN 28 for the augmenting stations of 0172, 0914 and 0918 during seismic shaking; b) Ionospheric interpolation errors of PRN 28 for the station 0175.*

Ohta et al. (2012). It can be seen that there are some fluctuations in the displacement series derived from the relative positioning solution, which are caused by the ground shaking at the reference station location. The Figure 3.2b shows the ground displacements at the 0065 reference station. Peak surface displacements of up to half a meter were recorded at this station during the earthquake even though it is about 700 km away from the epicenter. The displacement waveforms for the station 0588 derived from augmented PPP (0217, 0590 and 0965 are selected as augmenting stations), global PPP and RP solutions are also compared in Figure 3.6.

The permanent coseismic displacements of ninety evenly-distributed stations derived from post-processed ARIA solution (5 mins solution), real-time augmented PPP, global PPP, and RP solution are shown in Figures 3.7a, 3.7b, 3.7c, and 3.7d, respectively, by the red arrows. The post-processed ARIA solution is provided by the ARIA team at JPL (Jet Propulsion Laboratory) and Caltech (California Institute of Technology). It can be found that the permanent coseismic displacements, derived from the real-time augmented PPP solution, are quite consistent with those of post-processed ARIA solution in both horizontal and vertical components. The root mean squared errors (RMS) of the differences between the two solutions are 1.4 cm, 1.1 cm, and 1.7 cm in north, east, and vertical components, respectively. Figure 3.7c shows some significant differences between global PPP and ARIA displacements at some stations, which are caused by the data interruptions at these stations. The corresponding RMS values of the differences are 4.3 cm, 22.7 cm, and 9.0 cm in north, east, and vertical components. Figure 3.7d shows, that the RP displacements have obvious disagreements with the ARIA results at nearly all stations due to problem of the earthquake shaking of the reference station. The RMS values of the differences are 10.1 cm, 14.1 cm, and 5.7 cm in north, east, and vertical components. Figure 3.8 shows the displacement differences between the ARIA solution and the other three

Chapter 3 Augmented PPP for Seismological Applications Using Dense GNSS Networks

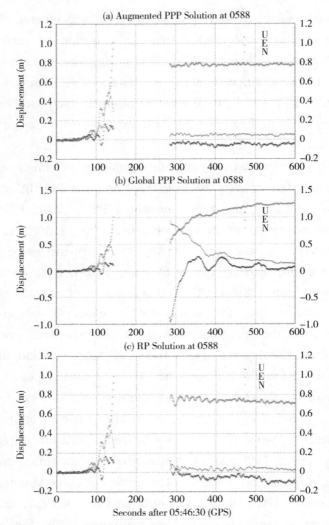

Figure 3.6 *Comparisons of displacement waveforms derived from augmented PPP, global PPP and RP solutions. The north, east and up components are shown by the blue, red and black curves, respectively. a) The displacements derived from augmented PPP solution at station 0588, which has a data gap of about 2 min during the seismic shaking; b) The displacements derived from global PPP solution at station 0588; c) The displacements derived from relative positioning at station 0588, with 0065 as the reference station.*

3.3 Application of Augmented PPP Approach and Results

solutions. These comparisons show that the augmented PPP method can significantly improve the reliability and accuracy of earthquake-induced coseismic displacements in real-time scenarios.

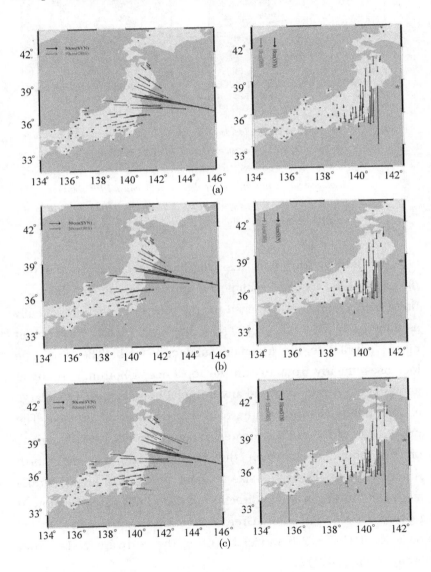

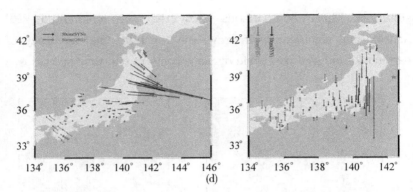

Figure 3.7 *The comparisons of observed and synthetic displacements in horizontal components, and in vertical components, respectively. (a) Inversion with the post-processed ARIA solution; (b) Inversion with coseismic displacements obtained from real-time augmented PPP solution; (c) Inversion with real-time global PPP solution; (d) Inversion with real-time RP solution. This figure is drawn using GMT software.*

We derived four fault slip distributions based on the four different GPS analysis techniques introduced above. Identical finite fault parameters are used for the four inversions. Identically as done by Wang et al. (2013), we employ a slightly curved fault plane, parallel to the assumed subduction slab. The dip angle increases linearly from 10° on the top (ocean bottom) to 20° at about 80 km depth. To avoid any artificial boundary effect, a large enough potential rupture area of 650 km × 300 km is used. The upper edge of the fault is located along the trench east of Japan, on the boundary between the Pacific plate and the North American plate. The patch size is about 10 km × 10 km. The rake angle determining the slip direction at each fault patch is allowed to vary between 90° ± 20°. Green's functions are calculated based on a local CRUST2.0 model by using the software codes from Wang et al. (2003).

3.3 Application of Augmented PPP Approach and Results

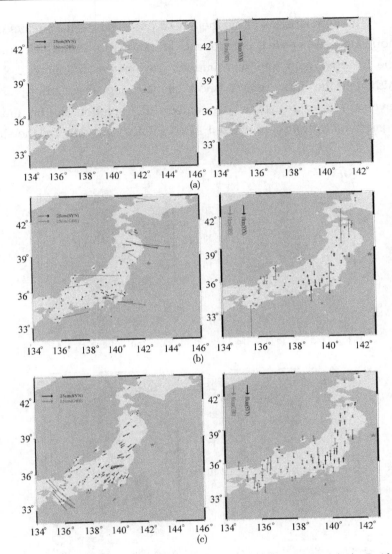

Figure 3.8 *The residual displacements from the ARIA solution. (a) Residual differences between augmented PPP and ARIA vectors; (b) Residual differences between global PPP and ARIA vectors; (c) Residual differences between RP and ARIA vectors. This figure is drawn using GMT software.*

The comparisons of synthetic and observed displacements on horizontal and vertical components are shown in Figure 3.7, and the inverted fault slip distributions are shown in Figure 3.9. Although the four results show similar slip distribution, the inversion from real-time augmented PPP solution is the most consistent with post-processed ARIA solution not only for the slip

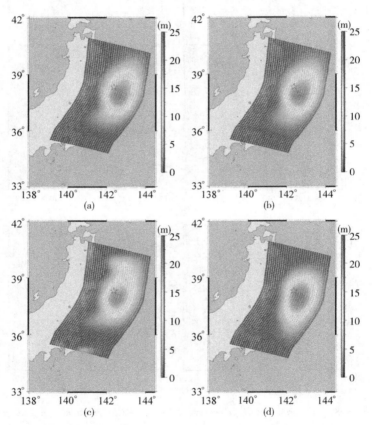

Figure 3.9 The inverted fault slip distributions. (a) Inversion with post-processed ARIA solution; (b) Inversion with coseismic displacements obtained from real-time augmented PPP solution; (c) Inversion with real-time global PPP solution; (d) Inversion with real-time RP solution. The star denotes the epicenter. This figure is drawn using GMT software.

distribution, but also for the displacement fittings. Supposing that the post-processed ARIA result is the most reliable and can be taken as a reference for other three results, the inversion of global PPP has the worst slip distribution, and the inversion of RP solution has the worst displacement fittings. Figure 3.10 shows the

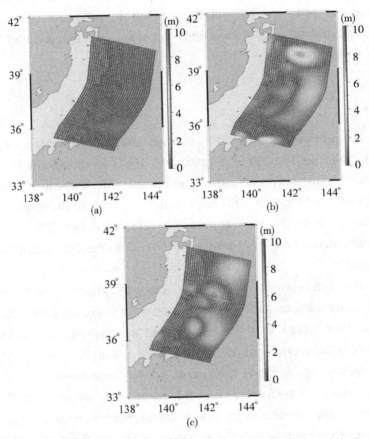

Figure 3.10 *The residual slip distributions from the ARIA solution. (a) Residual differences between augmented PPP and ARIA inversion; (b) Residual differences between global PPP and ARIA inversion; (c) Residual differences between RP and ARIA inversion. The star denotes the epicenter. This figure is drawn using GMT software.*

fault slip differences between the ARIA solution and the other three solutions. Overall, the comparison of the inversions shows that the augmented PPP method is beneficial for fault slip inversion in real-time scenarios. It provides a more accurate and robust estimation of the fault slip distribution and displacement fittings than the global PPP solution and RP solution. By contrast, the global PPP and RP solutions result in relatively poor slip distributions not only in peak slip, but also in the extension of the slip areas (see Figure 3.10).

3.4 Conclusions

We proposed a new GPS analysis method for hazard (e.g. earthquake and tsunami) monitoring. The new augmented PPP method can overcome the limitations of current relative positioning and global PPP approaches for this application. The performance of the new approach is evaluated by GPS ground network data, observed during the 2011 Tohoku-Oki earthquake in Japan.

The atmospheric corrections retrieved from the nearby monitoring stations can be interpolated with accuracy better than 5 cm. This means that the interpolated atmospheric corrections are accurate enough for rapid ambiguity resolution, which is a prerequisite to achieve the most precise displacements. The displacement waveforms, derived using the augmented PPP approach are immune to the convergence problem caused by data gaps and cycle slips and the problem of the earthquake shaking the reference station compared to the waveforms based on RP and global PPP analysis. This makes augmented PPP potentially appropriate for the application in operational earthquake/tsunami monitoring and warning systems. The reliability and accuracy of permanent coseismic displacements are also significantly improved. The RMS accuracy of about 1.4, 1.1, and 1.7 cms are achieved in the north, east, and vertical components, respectively. The inversion results indicate that the augmented PPP solution is the most consistent with post-processed ARIA solution both in the fault slip distribution and displacement fittings.

Chapter 4 Temporal Point Positioning Approach for GNSS Seismology Using a Single Receiver

4.1 Introduction

High-rate GNSS (e.g., 1 Hz or higher frequency) measures displacements directly and can provide reliable estimates of broadband displacements, including static offsets and dynamic motions of arbitrarily large magnitude (Larson et al., 2003; Bock et al., 2004). GNSS-derived displacements can be used to quickly estimate earthquake magnitude, model finite fault slip, and also play an important role in earthquake/tsunami early warning (Blewitt et al., 2006; Wright et al., 2012; Hoechner et al., 2013).

There are two primary strategies for real-time GNSS processing: relative baseline/network positioning and precise point positioning (PPP) (Zumberge et al., 1997). The disadvantage of relative positioning (RP) is that the solutions are influenced by movements of the reference stations (Ohta et al., 2012). Additionally the computational load increases very quickly as the network gets larger (Crowell et al., 2009). In contrast, PPP can provide "absolute" seismic displacements related to a global reference frame defined by the satellite orbits and clocks with a single GNSS receiver (Kouba 2003; Wright et al., 2012; Li et al., 2013a).

However, real-time PPP requires precise satellite orbits and clock corrections and also needs a long (re)convergence period, of about thirty minutes, to achieve centimeter-level accuracy (Collins et al., 2009).

Colosimo et al. (2011) proposed a variometric approach to overcome the difficulties of the two aforementioned, presently adopted, approaches for GNSS seismology. This approach is based upon the time single-differences of the carrier phase observations recorded by a single GNSS receiver at a given ground station. The time series of the station velocities are estimated, and these velocities are integrated to provide coseismic displacements. However, eventual biases of the estimated velocities accumulate over time and display as a drift in the coseismic displacements. The assumption of a linear drift limits the integration interval to few minutes (Branzanti et al., 2013). If the entire period of seismic shaking lasts longer than few minutes in the case of large earthquakes, the drift value could be large and can not be fully removed by a linear de-trending.

In this Chapter, we propose a new approach for estimating coseismic displacements with a single receiver in real-time. The approach overcomes not only the disadvantages of the PPP and RP techniques, but also decreases the described drift in the displacements derived from the variometric approach. The coseismic displacement could be estimated with few centimeters accuracy using GNSS data around the earthquake period. Meanwhile, we present and compare the observation models and processing strategies of the current existing single-receiver methods for real-time GNSS seismology. Furthermore, we propose several refinements to the variometric approach in order to eliminate the drift trend in the integrated coseismic displacements. The mathematical relationship between these methods is discussed in detail and their equivalence is also proved. The impact of error components such as satellite ephemeris, ionospheric delay, tropospheric delay, and geometry change on the retrieved displacements are carefully analyzed and investigated. Finally, the performance of these single-receiver approaches for real-time GNSS seismology is validated using 1 Hz GPS data collected during the Tohoku-Oki earthquake (Mw 9.0, March 11, 2011) in Japan.

4.2 Temporal Point Positioning Approach

The linearized equations for carrier phase and code observations can be expressed as follows,

$$l_{r,j}^s = -u_r^s \cdot x - t^s + t_r + B_{r,j}^s - I_{r,j}^s + T_r^s + \varepsilon_{r,j}^s \qquad (4.1)$$

$$p_{r,j}^s = -u_r^s \cdot x - t^s + t_r + I_{r,j}^s + T_r^s + e_{r,j}^s \qquad (4.2)$$

Where, $l_{r,j}^s$, $p_{r,j}^s$ denote "observed minus computed" phase and code observations from satellite s to receiver r at frequency j (j = 1, 2); u_r^s is unit direction vector from receiver to satellite; x denotes receiver position increments; T_r^s, $I_{r,j}^s$ denote tropospheric and ionospheric delay; t^s, t_r are clock errors of satellite and receiver; $B_{r,j}^s$ is phase ambiguity; $e_{r,j}^s$, $\varepsilon_{r,j}^s$ are measurement noise of carrier phase and code.

In order to achieve the most precise position estimates with GNSS, the phase center offsets and variations, and station displacements by tidal loading must be considered. Phase wind-up and relativistic delays must also be corrected according to the existing models (Kouba and Héroux, 2001), although they are not included in the equations. The ionospheric delays can be estimated as unknown parameters or eliminated by using dual-frequency phase and code data. The tropospheric delay is corrected with an a priori model, and the residual part is described as a random walk process (Boehm et al., 2006). The receiver clock is estimated epoch-wise as white noise. Furthermore, real-time precise satellite orbit and clock products are now available online via the International GNSS Service (IGS) real-time pilot project (RTPP) (Caissy et al., 2012; Dow et al., 2009).

For real-time PPP processing, the phase ambiguities are estimated together with the receiver position, receiver clock, and residual tropospheric delays. The ambiguities need some time to converge (e.g. thirty minutes) to the correct values, until enough observables are used in the filter. There will be a big disturbance in the displacement sequence during the convergence period (see Figure 4.1). In the variometric approach, ambiguities are eliminated using the time difference of phase observations and thus the convergence process is not required. Although the velocities can be estimated with a high accuracy, the integration

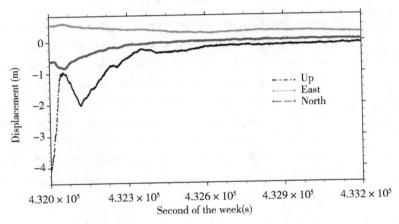

Figure 4.1 *PPP displacements during convergence period in the north, east and up components, respectively. Twenty minutes interval of 00:00-00:20 (GPST) on 11 March 2011, at GPS station 0177 (GEONET). The north, east, and up components are respectively shown by blue, red, and black lines.*

process from velocities to displacements may lead to accumulated drift if longer than few minutes (see results in Colosimo et al. 2011 and/or Figure 4.2).

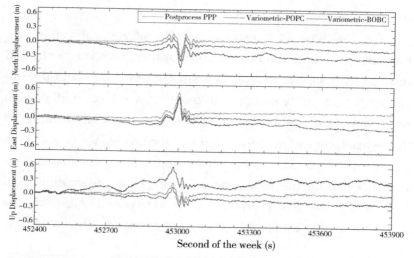

Figure 4.2 *Comparison of the coseismic displacement waveforms derived from the variometric approach and post-processed PPP solution. The red lines show the post-processed PPP waveforms as reference. The blue lines are the waveforms derived from the variometric approach using precise satellite orbits and clocks (POPC solution), while the black line indicates the result using broadcast orbits and clocks (BOBC solution). Twenty minutes interval around the entire period of seismic shaking on 11 March 2011, at GPS station 0986 (GEONET). From top to bottom are the results in north, east, and up components, respectively.*

In fact, for the seismological applications, we are mainly interested in the position variation relative to the position before the earthquake. Generally, the receiver position before the earthquake is well-known. Assuming that the receiver position at

4.2 Temporal Point Positioning Approach

the epoch t_0 (before the earthquake) is $x(t_0)$, the ambiguities $B(t_0)$ can be estimated along with the receiver clock $t_r(t_0)$ and tropospheric delay $T(t_0)$ (fixed to a priori model) parameters at this epoch as,

$$B(t_0) + t_r(t_0) + T(t_0) = l(t_0) + u(t_0) \cdot x(t_0) + t^s(t_0) - \varepsilon(t_0) \tag{4.3}$$

In our processing, all the error components are carefully considered following the PPP model. When the receiver position $x(t_0)$ is well known, the ambiguities $B(t_0)$ with a certain accuracy can be expected. Then we hold the estimated ambiguities $B(t_0)$ fixed in the subsequent epochs. At the epoch t_n, the positions $x(t_n)$ can be estimated as,

$$u(t_n) \cdot x(t_n) - t_r(t_n) - T(t_n) = -l(t_n) - t^s(t_n) + B(t_0) + \varepsilon(t_n) \tag{4.4}$$

As the ambiguities are held to fixed values instead of being estimated as unknown parameters, the convergence process will not be required. Furthermore, the positions $x(t_n)$ are estimated directly and thus the integration process is also avoided. We substitute the equation (4.3) into the equation (4.4) and have,

$$u(t_n) \cdot x(t_n) - \Delta t_r(t_0, t_n) + \Delta T(t_0, t_n)$$
$$= u(t_0) \cdot x(t_0) + \Delta l(t_0, t_n) + \Delta t^s(t_0, t_n) - \Delta \varepsilon(t_0, t_n) \tag{4.5}$$

It can be found that the accuracy of the position estimates $x(t_n)$ is mainly affected by the variation of the tropospheric delay from the epoch t_0 to t_n. Generally, the variation of the tropospheric delay is at centimeter level for few tens of minutes. Therefore, the position estimates are reasonably presumed to be with a good accuracy at centimeter level.

We can see that equation (4.5) is in the same form as the time-differenced equation of phase observations between the epoch t_0 and t_n. This is equivalent to calculating the position at epoch t_n relative to the well-known position at epoch t_0. This method is based on observations from a single receiver. Therefore, we refer to it as the temporal point positioning (TPP) method in the following sections. In our approach, an accurate initial position at epoch t_0 (i.e. the receiver position before the earthquake) is important for achieving high-accuracy displacements. Figure 4.3 shows the displacement results using initial position with different accuracies.

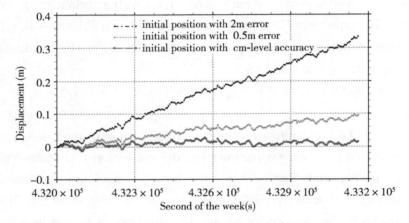

Figure 4.3 *Displacements derived from real-time TPP solution in east component. The blue line shows the result using initial position with cm-level accuracy, the red line is the result using initial position with 0.5m error, and the black line is the result using initial position with 2m error. Twenty minutes interval of 00:00-00:20 (GPST) on 11 March 2011, at GPS station 0986 (GEONET).*

4.3 Application of TPP Approach and Results

The 2011 Mw 9.0 Tohoku-Oki earthquake (11 March 2011, 05:46:23 UTC) in Japan is one of the best recorded large-

4.3 Application of TPP Approach and Results

magnitude earthquakes in history, by GNSS, as Japan has one of the most dense GNSS ground networks in the world. This network is operated by the Geospatial Information Authority of Japan (GSI) and consists of more than 1,200 continuously observing GNSS stations (the GNSS Earth Observation Network System, GEONET) all over Japan (http://www.gsi.go.jp/).

Firstly, the 1 Hz GEONET GPS data (dual frequency) before the earthquake was processed to evaluate the accuracy of the proposed TPP method. Twenty minutes of displacements from 00:00-00:20 (GPST) on 11 March 2011, at GNSS station 0986, are shown in Figure 4.4. We compare the displacements, derived from the real-time TPP solution using different orbit and clock products. The black lines show the results using broadcast orbit and clock (BOBC solution), which is routinely available from the GNSS receiver itself in real-time. The red lines are the results using precise satellite orbit and clock solutions (POPC solution). The satellite orbit is generally predicted for real-time applications as its dynamic stability. Here the ultra-rapid orbit, updated every three hours and provided by GFZ, is applied. The clock corrections have to be estimated and updated much more frequently (Zhang et al. 2011) due to their short-term fluctuation. We process 1 Hz data from 80-90 globally distributed real-time IGS stations using the GFZ's EPOS-RT software (Ge et al., 2011) in simulated real-time mode (a strictly forward filter) for generating precise GNSS clock corrections at a 5 s sampling interval.

Real-time orbit and clock corrections are reliant on an internet connection for transmission to monitoring stations, but the reality is that the internet connection and communication infrastructure could be destroyed during large earthquakes. In these cases, satellite clock corrections have to be extrapolated, although predicted ultra-rapid satellite orbits from the IGS or GFZ can be downloaded in advance. Here we also evaluate how our products would be degraded during large earthquake, when the real-time stream would be unavailable. The results using the precise predicted orbit and extrapolated clock (POEC solution) are shown by the blue line.

From Figure 4.4, we can see that there is no obvious drift for even a twenty minute period in the horizontal components of TPP derived displacements when precise orbit and clock corrections are applied. In the vertical component, there is only a small drift of a few centimeters. When only broadcast orbit and clock products are available, the drifts in the displacement series are clearly visible, especially in the vertical component. The drift values for twenty minutes are several centimeters in the horizontal components, and a few decimeters in the vertical component. If we compare this result (TPP with BOBC) and Figure 4.2 result (variometric approach with BOBC), we can see that the two approaches display similar accuracy level. In the scenario that we can only use extrapolated satellite clocks due to failure of the internet connection, the drift values are several centimeters in the horizontal components, and about one decimeter in the vertical component. This is several centimeters worse than the precise clocks results, but much better than the broadcast orbit and clock results, particularly in the vertical component. Obviously, the accuracy of orbits and clocks plays a crucial role, and the differences between POPC, POEC and BOBC are reduced to few centimeters if only the first 3-4 minutes are considered.

We calculated the root mean squares (RMS) of the drift errors at twenty minutes of eighty evenly-distributed GEONET stations. The results for the different orbit and clock products are summarized in Table 4.1. The drifts of the BOBC solution can reach up to 9.1, 7.8, and 28.2 cms in north, east and up

4.3 Application of TPP Approach and Results

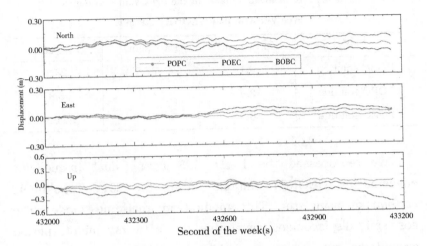

Figure 4.4   *Displacements derived from real-time TPP solution. The red line shows the result using precise satellite orbit and clock (POPC solution), the blue line is the result using precise orbit and extrapolated clock (POEC solution), and the black line is the result using broadcast orbit and clock (BOBC solution). Twenty minutes interval of 00:00-00:20 (GPST) on 11 March 2011, at GNSS station 0986 (GEONET). From top to bottom are the results in north, east, and up components, respectively.*

directions respectively. The precise orbit and clock corrections can remarkably improve the accuracy to 2.9, 2.3 and 5.8 cms in the corresponding directions. It is even comparable to the accuracy of PPP after convergence period. With the extrapolated satellite clocks, accuracies of 5.3, 4.7, 11.3 cms can be achieved in three components, respectively. Although these accuracies are degraded compared to the POPC solution, it significantly improves the BOBC solution. From these results, we also found that the vertical component is the most sensitive to the quality of the orbit and clock products.

Table 4.1 Root mean squares of the drift values at eighty evenly-distributed GEONET stations

RMS	North (cm)	East (cm)	Up (cm)
BOBC solution	9.1	7.8	28.2
POPC solution	2.9	2.3	5.8
POEC solution	5.3	4.7	11.3

We reprocessed the 1 Hz GPS data (dual frequency) collected by GEONET stations during the 2011 Tohoku-Oki earthquake using the TPP method in real-time mode. The coseismic displacement waveforms, for a twenty minute period around the entire seismic shaking at GNSS station 0986, are shown in Figure 4.5. The TPP waveforms using precise satellite orbits and clocks are shown by the blue line. The post-processed PPP waveforms, which have an accuracy of few centimeters (Kouba 2003; Wright et al., 2012), can be regarded as a reference, and are shown by the red line. The comparisons between them show that the TPP waveforms are quite consistent with the PPP results at a few centimeters accuracy during the entire shaking period. When only broadcast orbits and clocks are applied to the processing, the performance of the TPP method is degraded to about one decimeter in the horizontal components and about two decimeters in the vertical component, as indicated by the black line.

For comparison, we also process all the data using the variometric approach. All the error components are carefully corrected following the PPP and/or TPP model. The cumulative displacements at GNSS station 0986 are shown in Figure 4.2, illustrating typical behavior. Although precise orbits and clocks are applied, there are drifts up to few decimeters in the cumulative

4.3 Application of TPP Approach and Results

displacements. Compared to Figure 4.5, one can see that the TPP method can improve the displacement accuracy for POPC solution if the duration is longer than few minutes.

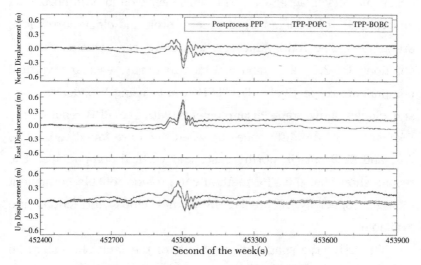

Figure 4.5 *Comparison of the coseismic displacement waveforms derived from real-time TPP solution and post-processed PPP solution. The red line shows the post-processed PPP result as a reference for TPP results, the blue line is the TPP result using precise satellite orbit and clock (POPC solution), and the black line is the result using broadcast orbit and clock (BOBC solution). A twenty minute interval around the entire period of seismic shaking on 11 March 2011, at GNSS station 0986 (GEONET) is shown. From top to bottom are the results in north, east, and up components, respectively.*

The permanent coseismic offset is the important information for magnitude estimation and fault slip inversion. In Figure 4.6, we compare the permanent coseismic offsets of eighty evenly-distributed stations derived from the TPP solution and the post-processed PPP solution. The post-processed PPP results, TPP

83

results using precise satellite orbit and clock, and TPP results using broadcast orbit and clock are shown respectively by the red, green and purple arrows. It is found that the permanent TPP coseismic offsets agree with PPP ones very well in both horizontal and vertical components when precise orbit and clock corrections are applied. The RMS values of the differences between the two solutions are 3.0, 2.1, and 5.6 cms in north, east and vertical components respectively. The TPP results using broadcast orbits and clocks show some disagreements with the PPP results; the RMS values of the differences between them are found to be 8.2, 7.0, and 22.9 cms in north, east, and vertical components. The results show that the TPP method can provide reliable permanent offsets, especially if precise orbit and clock corrections are available.

We derive the spatial distributions of the fault slip using the coseismic displacements obtained from the real-time TPP solution with broadcast orbit/clock, TPP with precise orbit/clock, and post-processed PPP solution, respectively. In the same way as done by Wang et al. (2013), we employ a slightly curved fault plane, parallel to the assumed subduction slab. The dip angle increases linearly from 10° on the top (ocean bottom) to 20° at about 80 km depth. To avoid any artificial bounding effect, a large potential rupture area of 650 × 300 km is used. The upper edge of the fault is located along the trench east of Japan, on the boundary between the Pacific plate and the Eurasian plate. The patch size is 10 × 10 km. The rake angle determining the slip direction at each fault patch is allowed to vary between 90°± 20°. Green's functions are calculated based on the CRUST2.0 model (Bassin et al., 2000) in the relevant area.

4.3 Application of TPP Approach and Results

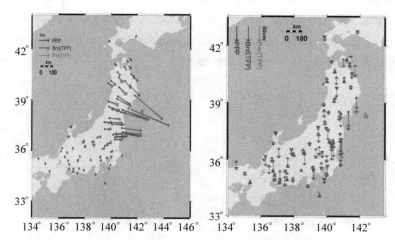

Figure 4.6 *A comparison of the permanent coseismic offsets derived from real-time TPP solution and post-processed PPP solution on horizontal components and on vertical components, respectively. The red arrow denotes the post-processed PPP result as a reference for TPP results, the green arrow denotes the TPP result using precise satellite orbit and clock, and the purple arrow is the result using broadcast orbit and clock.*

The three inversions result in moment magnitudes of Mw 8.90, 8.96, and 8.97, respectively. The maximum slip of the three inversion results are 21.0 m, 23.0 m and 23.3 m, respectively. The inverted fault slip distributions are shown in Figure 4.7. The post-processed PPP result is considered to be the most reliable and is taken as the reference. The inversion results of real-time TPP with precise orbits/clocks and the post-processed PPP solution are quite consistent with each other not only in the moment magnitude, but also in the slip distribution pattern. TPP using broadcast orbits/clocks leads to an underestimation of the moment magnitude and fault slip values to an extent. The comparison of the three inversion results shows that the TPP method can provide a reliable estimation of earthquake magnitude and of the fault slip distribution, especially when precise satellite orbit and clock corrections are used.

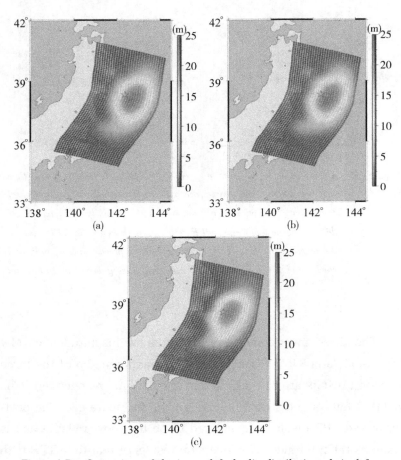

Figure 4.7 *Comparison of the inverted fault slip distributions derived from real-time TPP solution and post-processed PPP solution. From left to right are the inversion results derived from post-processed PPP, real-time TPP using precise orbit/clock, and real-time TPP using broadcast orbit/clock, respectively.*

4.4 Single-Receiver Approaches for Real-Time GNSS Seismology

4.4.1 Comparison of Analysis Methods

The linearized equations for undifferenced (UD) carrier phase and pseudorange observations can be respectively expressed as following,

$$l_{r,j}^s = -\mathbf{u}_r^s \cdot x + t_r - o^s - t^s - I_{r,j}^s + T_r^s + \lambda_j(N_{r,j}^s + b_{r,j} - b_j^s) + \varepsilon_{r,j}^s \tag{4.6}$$

$$p_{r,j}^s = -\mathbf{u}_r^s \cdot x + t_r - o^s - t^s + I_{r,j}^s + T_r^s + c(d_{r,j} + d_j^s) + e_{r,j}^s \tag{4.7}$$

where $l_{r,j}^s$ and $p_{r,j}^s$ denote "observed minus computed" phase and pseudorange observables from satellite s to receiver r at the frequency j; $\mathbf{u}_r^s$ is the unit vector of the direction from the receiver to the satellite; x denotes the vector of receiver position increments relative to a priori position x_0, which is used for linearization; o^s denotes satellite orbit error; t^s and t_r are clock errors of satellite and receiver respectively; $I_{r,j}^s$ is the ionospheric delay on the path at the j frequency; T_r^s denotes the tropospheric delay along the path; λ_j is the wavelength; $N_{r,j}^s$ is the integer phase ambiguity; $b_{r,j}$ and b_j^s are receiver-and satellite-dependent uncalibrated phase delays (UPD); $d_{r,j}$ and d_j^s are code biases of receiver and satellite; $e_{r,j}^s$ denotes pseudorange measurement noise and multipath; $\varepsilon_{r,j}^s$ denotes measurement noise of carrier phase and multipath.

In real-time PPP processing, the phase center offsets and variations and station displacements by tidal loading must be considered. Phase wind-up and relativistic delays must also be corrected according to the existing models (Kouba and Héroux, 2001), although they are not included in the equations. With the available real-time precise satellite orbit and clock products from the International GNSS Service (IGS) real-time pilot project (RTPP) (Caissy et al., 2012; Dow et al., 2009), the errors of satellite orbit and clock are greatly reduced to a few centimeters and can be neglected here. The ionospheric delays can be eliminated by the ionosphere-free linear combination (Kouba and Héroux, 2001) or can be processed by estimating the slant ionospheric delays in raw observations as unknown parameters (Li et al., 2013c). The tropospheric delay is corrected with an a priori model, and the residual part is estimated as a random walk process (Boehm et al., 2006). If UPD corrections are available, the UD ambiguities will have integer feature and can be fixed to integer values. Otherwise, the UD ambiguities are estimated as float values. A modified sidereal filtering proposed by Choi et al. (2004) could be used to mitigate the multipath error, but it is neglected here at present. The estimated parameters are,

$$X = (x^T \ t_r \ T_r (I_{r,1}^s)^T (N_{r,j}^s)^T)^T \quad (4.8)$$

A sequential least square or Kalman filter can be employed to estimate the unknown parameters for real-time processing. The increments of the receiver position x are estimated epoch by epoch without any constraints between epochs for retrieving rapid station movements. The receiver clock is estimated epoch-wise as white noise. The ionospheric delays are taken as estimated parameters for each satellite and at each epoch by using dual-frequency carrier phase and pseudorange observations. The residual tropospheric delay T_r is described as a random walk process with noise of about 2-5 $mm/\sqrt{hour}$. The carrier-phase ambiguities $N_{r,j}^s$ are estimated as constant over time until successful ambiguity fixing or convergence.

4.4 Single-Receiver Approaches for Real-Time GNSS Seismology

Colosimo et al. (2011) proposed a variometric approach for real-time GNSS seismology. This approach is based upon the time single-difference (SD) of the carrier phase observations recorded by a single GNSS receiver. The model of variometric approach can be derived from the time single-difference of UD observation equations (4.6) and (4.7) between two consecutive epochs (t_n, t_{n+1}) on the assumption that the observation data is continuous as follows,

$$\Delta l_{r,j}^s(t_n, t_{n+1}) = - \mathbf{u}_r^s(t_{n+1}) \cdot x(t_{n+1}) + \mathbf{u}_r^s(t_n) \cdot x(t_n)$$
$$+ \Delta t_r(t_n, t_{n+1}) - \Delta o^s(t_n, t_{n+1})$$
$$- \Delta t^s(t_n, t_{n+1}) - \Delta I_{r,j}^s(t_n, t_{n+1})$$
$$+ \Delta T_r^s(t_n, t_{n+1}) + \Delta \varepsilon_{r,j}^s(t_n, t_{n+1}) \quad (4.9)$$

$$\Delta p_{r,j}^s(t_n, t_{n+1}) = - \mathbf{u}_r^s(t_{n+1}) \cdot x(t_{n+1}) + \mathbf{u}_r^s(t_n) \cdot x(t_n)$$
$$+ \Delta t_r(t_n, t_{n+1}) - \Delta o^s(t_n, t_{n+1})$$
$$- \Delta t^s(t_n, t_{n+1}) + \Delta I_{r,j}^s(t_n, t_{n+1})$$
$$+ \Delta T_r^s(t_n, t_{n+1}) + \Delta e_{r,j}^s(t_n, t_{n+1}) \quad (4.10)$$

where $\Delta l_{r,j}^s(t_n, t_{n+1})$ is time single-difference phase observation $l_{r,j}^s(t_{n+1}) - l_{r,j}^s(t_n)$; $\Delta p_{r,j}^s(t_n, t_{n+1})$ is time single-difference pseudorange observation; $\mathbf{u}_r^s(t_n)$ and $\mathbf{u}_r^s(t_{n+1})$ are the unit direction vectors from receiver to satellite at epoch t_n and t_{n+1}; $x(t_n)$ and $x(t_{n+1})$ are the receiver position increments at epoch t_n and t_{n+1}; Other items represent the variation of the corresponding error components between epochs (t_n, t_{n+1}), for example, $\Delta I_{r,j}^s(t_n, t_{n+1})$, $\Delta T_r^s(t_n, t_{n+1})$ represent range variation caused by tropospheric and ionospheric refraction delay. Compared with the equations (4.6) and (4.7), it can be seen that phase ambiguities ($N_{r,j}^s$), phase delays ($b_{r,j}, b_j^s$) and code biases ($d_{r,j}, d_j^s$) can be eliminated through the time difference operation, as they can be regarded as constants for at least tens of minutes.

The accuracy of phase observation is much higher (about 100

times) than the pseudorange observation, thus the time-differenced position is mainly determined by phase observation. We will hereafter focus on phase observation; the equation (4.9) can be reformulated as,

$$\Delta l^s_{r,j}(t_n, t_{n+1}) = - \mathbf{u}^s_r(t_{n+1}) \cdot (x(t_{n+1}) - x(t_n)) - (\mathbf{u}^s_r(t_{n+1})$$
$$- \mathbf{u}^s_r(t_n)) \cdot x(t_n) + \Delta t_r(t_n, t_{n+1})$$
$$+ \Delta err^s_{r,j}(t_n, t_{n+1})$$
$$= - \mathbf{u}^s_r(t_{n+1}) \cdot \Delta x(t_n, t_{n+1}) - (\mathbf{u}^s_r(t_{n+1})$$
$$- \mathbf{u}^s_r(t_n)) \cdot x(t_n) + \Delta t_r(t_n, t_{n+1})$$
$$+ \Delta err^s_{r,j}(t_n, t_{n+1}) \qquad (4.11)$$

$$\Delta err^s_{r,j}(t_n, t_{n+1}) = - \Delta o^s(t_n, t_{n+1}) - \Delta t^s(t_n, t_{n+1}) - \Delta I^s_{r,j}(t_n, t_{n+1})$$
$$+ \Delta T^s_r(t_n, t_{n+1}) + \Delta \varepsilon^s_{r,j}(t_n, t_{n+1}) \qquad (4.12)$$

$\Delta x(t_n, t_{n+1})$ is the change in the receiver position increments for the time interval (t_n, t_{n+1}), which is the quantity of greatest interest; $\Delta t_r(t_n, t_{n+1})$ is the change in the receiver clock error; $\Delta err^s_{r,j}(t_n, t_{n+1})$ represent the sum of changes in all other error components; $(\mathbf{u}^s_r(t_{n+1}) - \mathbf{u}^s_r(t_n)) \cdot x(t_n)$ accounts for the change in the relative satellite/receiver geometry due to the line-of-sight vector changes its orientation. The estimated parameters are,

$$X = (\Delta x (t_n, t_{n+1})^T \Delta t_r(t_n, t_{n+1}))^T \qquad (4.13)$$

which can be easily estimated by using the least squares method when at least four satellites are being tracked simultaneously.

In the variometric approach, the velocities can be estimated with a high accuracy on the order of mm/s using a high-rate stand-alone receiver. A discrete integration of estimated velocities is then employed to reconstruct the coseismic displacement. Well-known is that this discrete integration is very sensitive to estimation biases due to a possible mismodeling of different intervening effects that accumulate over time and display their signature as a trend in coseismic displacements. The trend can be assumed to be linear if the integration interval was limited up to few minutes (Branzanti et al., 2013). Furthermore, the variometric approach is effective even when using a simplified model with broadcast orbit and clock and single frequency receiver, which the effects due to the ionosphere, troposphere, phase center variation, relativity, and phase wind-up are neglected (Colosimo et al. 2011).

In order to eliminate or significantly decrease the drift trend in the integrated displacements and also avoid the linear detrending process, we propose several refinements to the variometric approach: 1) Variations in satellite orbit error $\Delta o^s(t_n, t_{n+1})$ and clock bias $\Delta t^s(t_n, t_{n+1})$ are corrected by using real-time precise satellite orbit and clock products which are now available online via the IGS RTPP (Caissy et al., 2012). 2) All of the other error components are carefully corrected following the PPP model. Ionospheric delay changes are compensated using dual frequency measurements. The changes in tropospheric delay is mostly mitigated by a priori tropospheric model (Saastamoinen, 1972), the residual part is at centimeter level for few tens of minutes. The changes in the phase center offsets and variations, tidal loading, phase wind-up and relativistic delays can be corrected according to the existing models. 3) Special attention is given to the geometry correction ($\mathbf{u}_r^s(t_{n+1}) - \mathbf{u}_r^s(t_n)) \cdot x(t_n)$, which accounts for changes in the relative satellite/receiver geometry.

Usually, the geometry error item is ignored or an approximate receiver position estimated from standard point positioning (SPP) is used to calculate and correct it. However, the geometry error could be large if the integration duration is longer than few minutes (the line-of-sight vector change $\mathbf{u}_r^s(t_{n+1}) - \mathbf{u}_r^s(t_0)$ will be large) or the approximate SPP position is not accurate enough (the error of $x(t_n)$ will be large). In seismological applications, we are mainly interested in the displacements relative to the receiver position before the earthquake and the position before the earthquake is generally well-known. This accurate receiver position can be used to fully correct the geometry error. Assuming that the receiver position before the earthquake $x(t_0)$ is accurately known, it can be used to correct the geometry error ($\mathbf{u}_r^s(t_1) - \mathbf{u}_r^s(t_0)$) $\cdot x(t_0)$ in $\Delta x(t_0, t_1)$ estimation. Then we can have,

$$x(t_1) = x(t_0) + \Delta x(t_0, t_1) \quad (4.14)$$

By analogy, the integrated $x(t_1)$ can be used to correct the geometry error ($\mathbf{u}_r^s(t_2) - \mathbf{u}_r^s(t_1)$) $\cdot x(t_1)$ in $\Delta x(t_1, t_2)$ estimation, and so on. The integrated $x(t_n)$ is used to correct the geometry error ($\mathbf{u}_r^s(t_{n+1}) - \mathbf{u}_r^s(t_n)$) $\cdot x(t_n)$ in $\Delta x(t_n, t_{n+1})$ estimation.

After all of error sources are carefully considered, the integrated displacements from the refined variometric approach are reasonably presumed to be with a good accuracy at centimeter level without the need of de-trending. The accuracy mainly depends on the variation of residual tropospheric delay, which is at centimeter level for few tens of minutes.

We recently proposed an innovative TPP approach to single point, single epoch, GNSS positioning at few centimeters precision level over a period up to about 20 minutes (Li et al., 2013b), which is typically required for coseismic displacement determinations after major earthquakes. Based on the facts that: 1) the position change (relative to the position before the earthquake) is the quantity of greatest interest in seismological applications; 2) the receiver position before the earthquake is generally well known. The model of TPP approach can be derived from UD observation equations (4.6) and (4.7) as follows (phase observation is concentrated here as its much higher precision),

$$B(t_0) + t_r(t_0) + T_r^s(t_0) = l_{r,j}^s(t_0) + \mathbf{u}_r^s(t_0) \cdot x(t_0) + o^s(t_0)$$

$$+ t^s(t_0) + I^s_{r,j}(t_0) - \varepsilon^s_{r,j}(t_0) \quad (4.15)$$
$$B(t_0) = \lambda_j(N^s_{r,j} + b_{r,j} - b^s_j) \quad (4.16)$$

The receiver position at the epoch t_0 (before the earthquake) is assumed to be precisely known as $x(t_0)$. All the error components including satellite orbit and clock errors, ionospheric and tropospheric delays, phase center offsets and variations, tidal loading, phase wind-up and relativistic delays are carefully considered following the PPP model. The real-valued ambiguities $B(t_0)$ can be estimated along with the receiver clock $t_r(t_0)$ and tropospheric delay $T(t_0)$ (tightly constrained or fixed to a priori model) parameters at this epoch. Then we hold the estimated ambiguities $B(t_0)$ fixed in the subsequent epochs. At the epoch t_n, the positions $x(t_n)$ can be estimated as,

$$\mathbf{u}^s_r(t_n) \cdot x(t_n) - t_r(t_n) - T^s_r(t_n) =$$
$$- l^s_{r,j}(t_n) - t^s(t_n) - o^s(t_n) + B(t_0) - I^s_{r,j}(t_n) + \varepsilon^s_{r,j}(t_n)$$
$$(4.17)$$

As the ambiguities are held to fixed values instead of being estimated as unknown parameters, the convergence process will not be required. Furthermore, the positions $x(t_n)$ are estimated directly and thus the integration process is also avoided.

When the receiver position $x(t_0)$ is precisely known and all of error sources are carefully considered, the ambiguities $B(t_0)$ with a certain accuracy can be expected. The accuracy of the position estimates $x(t_n)$ will be mainly affected by the variation of residual tropospheric delay from the epoch t_0 to t_n. Generally, the variation of the tropospheric delay is at centimeter level for few tens of minutes. Therefore, the position estimates are reasonably presumed to be with a good accuracy at centimeter level without de-trending process.

In TPP approach, an accurate initial position at epoch t_0 (i.e. the receiver position before the earthquake) is important for achieving high-accuracy displacements. The influence of initial position accuracy on the TPP displacements can be found in the section 4.2.

For real-time PPP processing, the phase ambiguities are estimated together with the receiver position, receiver clock, and residual tropospheric delays. The ambiguities need some time to converge (e.g. thirty minutes) to the correct values, until enough observables are used in the filter. There will be a big disturbance in the displacement sequence during the (re)convergence period. Once the ambiguities are successfully fixed to correct integer values or converged to accurate real values, displacement accuracy of few centimeters can be achieved.

In order to avoid the convergence problem, the TPP method makes full use of two critical features in seismological applications: the receiver position before the earthquake is generally well known and the ambiguity is constant on the assumption that the observation data is continuous. The TPP method is equivalent to PPP with the real phase ambiguities fixed at values determined from the known position at the epoch, preceding the earthquake and using an a priori tropospheric delay. This new TPP method can provide about the same, a cm level precision, as the converged PPP, which requires up to 30 mins data prior the earthquake for a PPP solution convergence. It can be found that the TPP and the converged PPP have similar mathematical model, the difference between them is that TPP uses the known position at initial epoch to calculate accurate phase ambiguity, while PPP uses a period of observation data for phase ambiguity convergence.

In the model of TPP approach, we substitute the equation (4.15) into the equation (4.17) and then have,

$$\mathbf{u}_r^s(t_n) \cdot x(t_n) - \Delta t_r(t_0, t_n)$$
$$= \mathbf{u}_r^s(t_0) \cdot x(t_0) - \Delta l_{r,j}^s(t_0, t_n) + \Delta err_{r,j}^s(t_0, t_n) \quad (4.18)$$

4.4 Single-Receiver Approaches for Real-Time GNSS Seismology

$$\Delta err^s_{r,j}(t_0,t_n) = -\Delta o^s(t_0,t_n) - \Delta t^s(t_0,t_n) - \Delta I^s_{r,j}(t_0,t_n)$$
$$+ \Delta T^s_r(t_0,t_n) + \Delta \varepsilon^s_{r,j}(t_0,t_n) \quad (4.19)$$

Meanwhile, the equation (4.11) of variometric approach can be reformulated as,

$$\mathbf{u}^s_r(t_{n+1}) \cdot x(t_{n+1}) - \Delta t_r(t_n,t_{n+1})$$
$$= \mathbf{u}^s_r(t_n) \cdot x(t_n) - \Delta l^s_{r,j}(t_n,t_{n+1}) + \Delta err^s_{r,j}(t_n,t_{n+1}) \quad (4.20)$$

We can see that equation (4.18) derived from TPP model is in the same form as the time-differenced equation (4.20) in the refined variometric model. The difference is that the TPP method is equivalent to calculating the displacement at epoch t_n relative to the well-known position at epoch t_0, while the variometric approach uses time-differenced phase observations between two adjacent epochs t_n and t_{n+1} to calculate velocities. In the variometric approach, ambiguities are eliminated by using time difference operation, and thus the convergence process is also not required. But an integration process is needed to reconstruct displacements from velocities. For the time series from epoch t_0 to t_n, the equation (4.20) can be expressed as,

$$\mathbf{u}^s_r(t_1) \cdot x(t_1) - \Delta t_r(t_0,t_1) = \mathbf{u}^s_r(t_0) \cdot x(t_0) - \Delta l^s_{r,j}(t_0,t_1)$$
$$+ \Delta err^s_{r,j}(t_0,t_1)$$
$$\mathbf{u}^s_r(t_2) \cdot x(t_2) - \Delta t_r(t_1,t_2) = \mathbf{u}^s_r(t_1) \cdot x(t_1) - \Delta l^s_{r,j}(t_1,t_2)$$
$$+ \Delta err^s_{r,j}(t_1,t_2)$$
$$\vdots$$
$$\mathbf{u}^s_r(t_n) \cdot x(t_n) - \Delta t_r(t_{n-1},t_n) = \mathbf{u}^s_r(t_{n-1}) \cdot x(t_{n-1}) - \Delta l^s_{r,j}(t_{n-1},t_n)$$
$$+ \Delta err^s_{r,j}(t_{n-1},t_n) \quad (4.21)$$

The cumulative sum of equations (4.21) is,

$$\mathbf{u}^s_r(t_n) \cdot x(t_n) - \Delta t_r(t_0,t_n)$$
$$= \mathbf{u}^s_r(t_0) \cdot x(t_0) - \Delta l^s_{r,j}(t_0,t_n) + \Delta err^s_{r,j}(t_0,t_n) \quad (4.22)$$

The accumulated equation (4.22) is the same as the equation (4.18) derived from TPP. It means that the TPP and refined variometric approaches can be equivalent after all of error sources, especially orbit, clock and geometry errors, are carefully considered.

In PPP approach, once ambiguities are fixed or converged to correct values, it also has the same observation model as TPP method for subsequent epochs and the same equation as (4.18) can also be derived from PPP model. The only difference is that PPP can use a better tropospheric delay estimated from convergence process, while TPP and variometric approaches use a priori tropospheric delay model. Fortunately, the variation of the tropospheric delay is slow, at centimeter level for few tens of minutes, which is of greatest interest in seismological applications.

From the above-mentioned analysis on the single-receiver approaches for real-time GNSS seismology, we can conclude that the TPP, converged PPP and refined variometric approaches have equivalent mathematical model and should provide about the same displacement precision.

4.4.2 Error Analysis and Precision Validations

Numerous studies show that PPP is a powerful technique for seismological applications, and PPP-derived displacement accuracy is comparable to relative positioning method if undifferenced ambiguities are successfully fixed (Li et al., 2013a; Geng et al., 2013). Compared with PPP approach, there is still little research on detailed error analysis and precision validation of variometric and TPP approaches for coseismic displacement retrieving. In this section, we carefully analyze the impact of error components on variometric and TPP-derived displacements. The precision of these single-receiver approaches is also evaluated and compared by using 1 Hz GEONET (the GNSS Earth Observation Network System) data collected during the 2011 Mw 9.0 Tohoku-Oki earthquake in Japan. This network is operated by the Geospatial Information Authority of Japan (GSI) and consists of more than 1,200 continuously observing GNSS stations all over Japan (http://www.gsi.go.jp/).

The variometric approach firstly computes the delta position of one station between two adjacent epochs, and then the displacement waveform is reconstructed through the discrete integration method. Well-known is that this (discrete) integration is very sensitive to estimation biases due to the possible mismodeling of different intervening effects (such as orbit and clock errors, atmospheric errors and geometry errors) that accumulate over time and display their signature as a trend in the coseismic displacements. We design four schemes as listed in the following table to evaluate the impacts of these errors on cumulative displacements. All of the other error components (e.g. phase center offsets and variations, tidal loading, phase wind-up and relativistic delays) are carefully corrected according to the existing models as PPP (Kouba and Héroux, 2001).

Table 4.2 Four different schemes for the variometric approach

	Satellite ephemeris	Ionospheric delay	Geometry error
Scheme 1 (BOBC L1)	broadcast ephemeris	L1 observation	an approximate position
Scheme 2 (BOBC LC)	broadcast ephemeris	LC observation*	an approximate position
Scheme 3 (POPC)	precise orbit and clock	LC observation	an approximate position
Scheme 4 (POPC Geometry)	precise orbit and clock	LC observation	an accurate position

* LC: Ionosphere-Free Combination

4.4.2.1 Ionospheric Effect

The cumulative displacements from 00:00-00:20 (GPST, before the earthquake) on 11 March 2011 for station 0177 are exemplarily shown in Figure 4.8. The results using broadcast orbit, broadcast clock and L1 carrier phase observation (BOBC L1, ionospheric delay is not compensated) are depicted by the blue lines, and the results in which ionospheric delay is eliminated by using ionosphere-free combination observation (BOBC LC) are shown by the red lines.

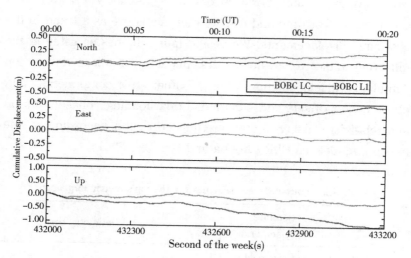

Figure 4.8 *Cumulative displacements using the variometric approach: impact of ionospheric delay. Results of station 0177 (GEONET) for 20 minutes interval 00:00:00-00:20:00 (GPST) on March 11, 2011.*

To clearly display the impact of ionospheric delay on the cumulative displacements, the displacement differences between BOBC L1 and BOBC LC results are shown in the blue line in Figure 4.9. A drift trend can be clearly seen in all three components. The linear fitting of displacement differences is also depicted in Figure 4.9 with the red line. Besides linear trend, the displacement differences also contain some short-term fluctuations because of the ionospheric disturbance. For 20 minutes integration interval, the displacement differences of station 0177 due to ionospheric delay could reach about 17 cm, 58 cm, and 83 cm in the north, east, and up components, respectively.

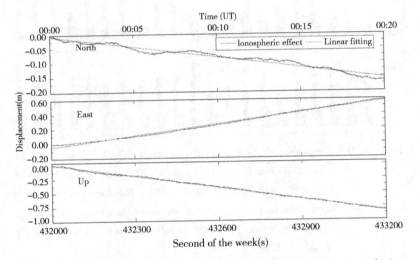

Figure 4.9 *The displacement differences of station 0177 due to ionospheric delay for the 20 minutes interval 00:00:00-00:20:00 (GPST) on March 11, 2011.*

The displacement errors of BOBC L1 and BOBC LC solutions of about twenty stations for 20 minutes interval are shown in Figure 4.10. The displacement errors of other stations are similar to the results of station 0177. The displacement errors of BOBC L1 solution are obviously larger than the BOBC LC solution, especially in up component.

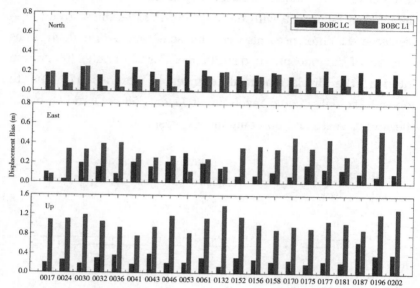

Figure 4.10 *Comparison of the displacement errors of BOBC L1 and BOBC LC solutions of about twenty stations for the 20 minutes interval 00:00:00-00:20:00 (GPST) on March 11, 2011. The displacement errors of north, east and up components are shown in the top, middle and bottom sub-figures. The BOBC L1 solutions are in red and the BOBC LC ones in blue.*

4.4.2.2 The Effect of Satellite Ephemeris

Currently, two types of orbit and clock products are available in real time. One is broadcast orbit and clock, which is routinely available from the GNSS receiver itself with an accuracy of about decimeter to meter level. The other one is precise satellite orbit and clock products from the International GNSS Service (IGS) real-time pilot project (RTPP) with an accuracy of few centimeters (Caissy et al., 2012; Dow et al., 2009). Here the

4.4 Single-Receiver Approaches for Real-Time GNSS Seismology

ultra-rapid orbit, updated every three hours and provided by GFZ, is applied. The clock corrections have to be estimated and updated much more frequently (Zhang et al. 2011) due to their short-term fluctuation. We process 1 Hz data from 80-90 globally distributed real-time IGS stations using the GFZ's EPOS-RT software (Ge et al., 2011) in simulated real-time mode (a strictly forward filter) for generating precise GNSS clock corrections at a 5 s sampling interval.

The cumulative displacements of station 0177 for 20 minutes interval are shown in Figure 4.11. The results using broadcast orbit, broadcast clock and LC observation (BOBC) are depicted by blue lines, and the results using precise orbit, precise clock and LC observation (POPC) are shown by red lines. Compared

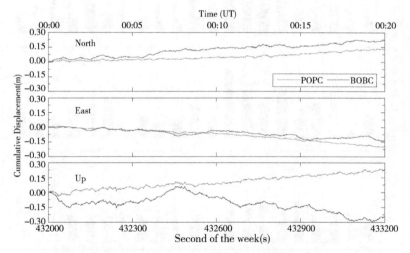

Figure 4.11 *Cumulative displacements using the variometric approach: impact of satellite orbit and clock. The red line shows the result using precise satellite orbit and clock, while the blue line is the result using broadcast clock and orbit. Results of station 0177 (GEONET) for the 20 minutes interval 00:00:00-00:20:00 (GPST) on March 11, 2011.*

with POPC results, BOBC results show a complicated drift character with more fluctuations. Take the up component for instance, the BOBC cumulative displacements has a drift up to about 30 cm.

The displacement errors of POPC and BOBC solutions of about twenty stations for 20 minutes interval are shown in Figure 4.12. The differences of the cumulative displacements between the two solutions denote the effect of satellite ephemeris on displacements, and displacement differences of the station 0177 could reach about 10 cm in horizontal components and about

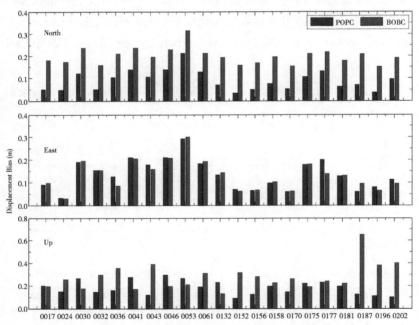

Figure 4.12 *Comparison of the displacement errors of POPC and BOBC solutions of about twenty stations for 20 minutes interval 00:00:00-00:20:00 (GPST) on March 11, 2011. The displacement errors in north, east and up components are shown in the top, middle and bottom sub-figures. The BOBC solutions are in red and the POPC ones in blue.*

50 cm in up component for 20 minutes integration interval. It can be found that the displacement errors of POPC solution are generally smaller than BOBC solution. In addition, the displacement differences are not linearly proportional to the integration interval. Taking the up component for instance, the differences between POPC and BOBC solutions on average are about 44.6 cm for 20 minutes interval, 14.9 cm for 10 minutes interval and 13.8 cm for 5 minutes interval. It is concluded that satellite ephemeris has important influence on accumulative displacements, and the displacement errors caused by broadcast orbit and clock is not a simple linear trend.

4.4.2.3 The Geometry Error Effect

The different cumulative displacement waveforms, devoted to the effect of geometry error for station 0177, are exemplarily shown in Figure 4.13. The result without compensating the geometry error by using precise ephemeris and LC observation (POPC) is depicted by the blue line, and the result in which geometry error is carefully corrected (POPC geometry) is shown by the red line. The differences between the two accumulative displacements indicate the effect of geometry error on the displacement. After 20 minutes, the displacement differences between POPC geometry results and POPC (no geometry correction) results could reach 13.3 cm, −20.3 cm, and 18.2 cm in the north, east, and up components, respectively. The POPC geometry solution, considered to be the most accurate estimation in four strategies, can achieve an accuracy of about few centimeters, which could be caused by the residual tropospheric delay.

The displacement errors of POPC geometry and POPC (no geometry correction) solutions of about twenty stations for 20 minutes interval are shown in Figure 4.14. It is obvious that the geometry error has very significant impact on accumulative displacements, it should be carefully considered to retrieve precise coseismic displacements. The displacement error caused by geometry item can reach up few decimeters for 20 minutes interval.

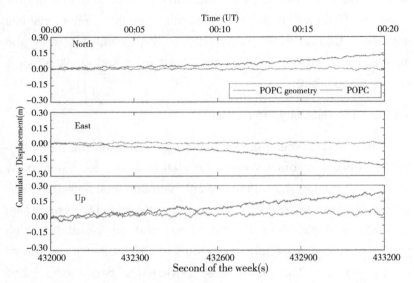

Figure 4.13 *Cumulative displacement waveforms using the integrated velocities: impact of the geometry error with (red line) and without (blue line) consideration. Results of 0177 (GEONET) in the 20 minutes interval 00:00:00-00:20:00 (GPST) on March 11, 2011.*

To further reflect the spectral characteristics of different displacement waveforms, the power spectral densities (PSD) at station 0177 for each scheme are shown in Figure 4.15. Four displacement results are respectively depicted by different color lines: (1) POPC geometry result in red line; (2) POPC result in blue line; (3) BOBC result in black line; and (4) BOBC L1 result

4.4 Single-Receiver Approaches for Real-Time GNSS Seismology

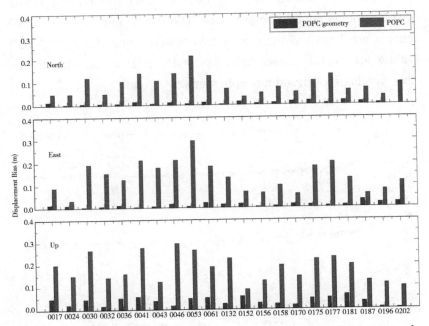

Figure 4.14 *Comparison of the displacement errors from POPC geometry and POPC solutions of about twenty stations for 20 minutes interval 00:00:00-00:20:00 (GPST) on March 11, 2011. The displacement errors in the north, east and up components are shown in the top, middle and bottom sub-figures. The POPC solutions are in red and the POPC geometry ones in blue.*

in cyan line. On the whole, the POPC geometry PSD curve performs more or less flat, especially at high frequency bands between 0.05 Hz and 0.5 Hz, but other three curves have many fluctuations mainly caused by their displacement waveforms with a nearly linear trend. At the low frequency bands less than 0.05 Hz, the POPC geometry PSD values are obviously the smallest in all three components, which indicates the POPC geometry displacements have few biases off the truth. Conversely, the

BOBC L1 PSD values are the biggest, and the corresponding displacements contain large biases. The POPC and the BOBC PSD curves are between the above two results, and the BOBC PSD values are slightly larger than the POPC PSD values due to the low precision of broadcast ephemeris.

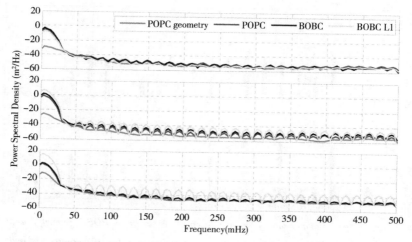

Figure 4.15 *Power spectral density comparisons of the displacement waveforms at station 0177 for each scheme: upper panel for the east component; middle panel for the north component; and lower panel for the vertical component.*

As ionospheric delay can be compensated by using dual-frequency observations and TPP method does not suffer from geometry error, we mainly concentrate on the impact of orbit and clock errors on the TPP displacements.

To investigate the effect of the satellite orbit and clock product on the TPP method, we process the 1 Hz GEONET data using different orbit and clock products. The TPP displacements of station 0177 for 20 minutes interval are shown in Figure 4.16.

4.4 Single-Receiver Approaches for Real-Time GNSS Seismology

The results using precise orbit and clock (POPC) is the closest to the zero line without a drift trend at an accuracy of few centimeters. The results using precise clock and broadcast orbit (PCBO) is the second closest to the zero line, but it gradually diverges from the zero value, especially in the up component. Compared with the POPC/PCBO results, the results using broadcast clock (both POBC and BOBC solutions) have an evident drift error up to few decimeters.

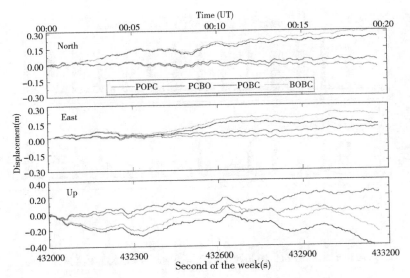

Figure 4.16 *Displacements using the TPP method: impact of satellite orbit and clock. Results of 0177 (GEONET) for 20 minutes interval 00:00:00-00:20:00 (GPST) on March 11, 2011.*

The TPP displacement errors of POPC, PCBO, POBC and BOBC solutions of about fifteen stations for 20 minutes interval are shown in Figure 4.17. It is clearly shown that the displacement errors for PCBO, POBC and BOBC solutions increases evidently up to few decimeters along with the extension of processing period, while the errors of POPC solution is only few centimeters in all three components. In addition, the displacements using precise satellite clock are much better than the ones using broadcast clock, both of the POBC and BOBC errors exceed one decimeter even when the integration interval is 5 minutes, while the POPC and PCBO errors are smaller than five centimeters. It can be concluded that the satellite clock error has more influence on the TPP results than satellite orbit error.

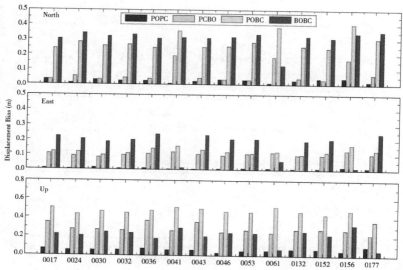

Figure 4.17 *Comparison of displacement errors from POPC, PCBO, POBC and BOBC solutions of about fifteen stations for 20 minutes interval 00:00:00-00:20:00 (GPST) on March 11 2011. The displacement errors in the north, east and up components are shown in the top, middle and bottom subfigures, respectively.*

The precise satellite clock product is accurate enough as a reference value, thus the difference between the broadcast and the precise clock product can be considered as error of broadcast satellite clock product. Figure 4.18 shows the broadcast clock error for satellite PRN 09 (see Figure S8 in auxiliary material for satellite PRN 10). We can find that the variation of clock errors could reach few decimeters for 20 minutes. The corresponding residual errors after a linear trend removal are also shown in bottom sub-figures.

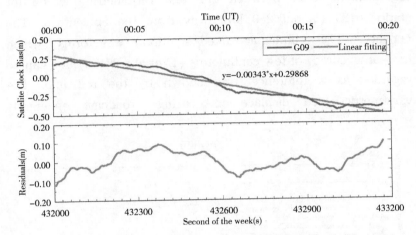

Figure 4.18 *The broadcast satellite clock error of PRN 09 for 20 minutes interval. The linear fitting results for the clock errors and the residuals after a linear trend removal are also shown in red line.*

4.4.3 Application to the 2011 Tohoku-Oki Earthquake

We reprocessed the 1 Hz GPS data (dual frequency) collected by GEONET stations during the 2011 Mw 9.0 Tohoku-Oki earthquake (11 March, 2011, 05:46:24 UTC; GPS Time-UTC = 15s) using single-receiver approaches (PPP, variometric and TPP) in real-time mode. For the PPP method, we processed these data using precise satellite orbit and clock product. For the variometric and TPP method, we processed these data using precise and broadcast orbit/clock products, respectively.

The coseismic displacement waveforms, for the twenty minute period around the entire seismic shaking at two GNSS stations (0183, 0986), are shown as examples from Figure 4.19 to Figure 4.22. The earthquake signature can be clearly observed in the 3D displacement waveforms. Figure 4.19 shows a comparison of displacement series between PPP and variometric method for station 0183 (about 250 km away from the epicenter). The converged PPP (ambiguity fixed solution) waveforms, which have an accuracy of few centimeters (Li et al., 2013a) and can be regarded as a reference, are shown by the red line. The variometric-based displacements using broadcast ephemeris

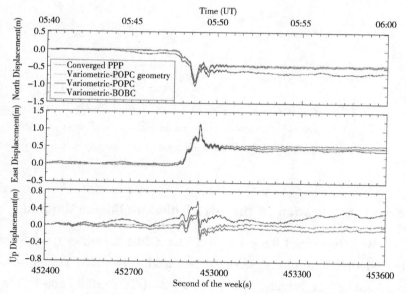

Figure 4.19 *Comparisons of the displacement waveforms using variometric method and converged PPP solution for station 0183 in the 20 minutes interval from 05:40:00 to 06:00:00 (GPST) on 11 March, 2011.*

4.4 Single-Receiver Approaches for Real-Time GNSS Seismology

(Variometric-BOBC) has a visible drift from the converged PPP results. Although precise orbits and clocks are applied, Variometric-POPC solution still drifts up to few decimeters in the cumulative displacements. After the geometry error correction, the variometric-based displacements (Variometric-POPC geometry) agree quite well with the converged PPP results.

Figure 4.20 shows a comparison between displacement waveforms derived from PPP and TPP method. The comparisons between them show that the TPP waveforms are quite consistent with the PPP results at few centimeters accuracy during the entire shaking period. The differences between the PPP (the red line) and TPP-POPC solution (the blue line) is within 3.0 cm in

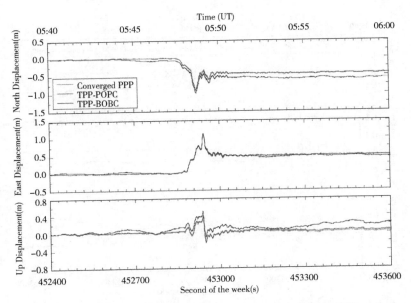

Figure 4.20 *Comparisons of the displacement waveforms using TPP method and converged PPP solution for station 0183 in the 20 minutes interval from 05:40:00 to 06:00:00 (GPST) on 11 March, 2011.*

horizontal components and within 5.0 cm in up component. When only broadcast orbit and clock are applied to the processing, the performance of the TPP method is degraded to about one decimeter in the horizontal components and about two decimeters in the vertical component, as indicated by the black line.

Figure 4.21 and Figure 4.22 are the displacements waveforms of station 0986 (about 485 km away from the epicenter). In view of the results of station 0986, we come to a similar conclusion like station 0183. When real-time precise orbit and clock corrections are available, TPP-POPC and Variometric-POPC geometry derive the displacement waveforms, which are both at a comparable level with the converged PPP waveforms at an

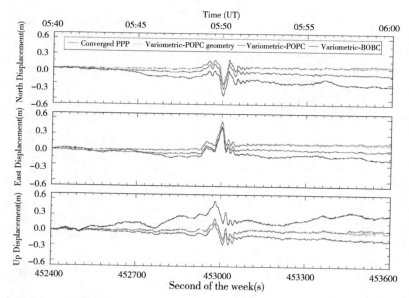

Figure 4.21 *Comparisons of the displacement waveforms using variometric method and converged PPP solution for station 0986 in the 20 minutes interval from 05:40:00 to 06:00:00 (GPST) on 11 March, 2011.*

accuracy of few centimeters during the entire shaking period, even for a period of twenty minutes.

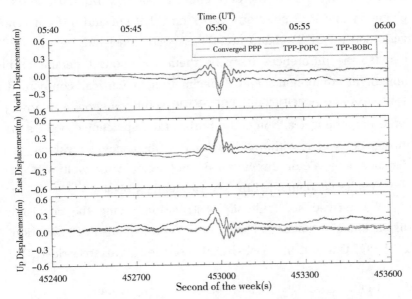

Figure 4.22 *Comparisons of the displacement waveforms using TPP method and converged PPP solution for station 0986 in the 20 minutes interval from 05:40:00 to 06:00:00 (GPST) on 11 March, 2011.*

The permanent coseismic displacements of ninety evenly-distributed stations derived from the PPP, TPP and variometric approach are shown in Figure 4.23. The PPP solution, which has been validated by numerous studies, is depicted as a reference here. In all the schemes, the TPP-POPC and variometric-POPC geometry solutions can achieve the most accurate coseismic displacements of about few centimeters (with a centralized direction to the earthquake source centroid), which agree quite well with the PPP results. The RMS of the differences between

TPP-POPC and PPP solutions is about 3.0 cm, 1.8 cm and 6.0 cm in the north, east and up components, respectively. The corresponding RMS of the differences between variometric-POPC-geometry and PPP solutions is 3.1 cm, 1.9 cm and 6.0 cm. The variometric-POPC results are the second consistent to the PPP results, the differences between them are about 1 decimeter in horizontal components and 2 decimeters in vertical component. The variometric-BOBC and TPP-BOBC solutions show a relatively large uncertainty, although the horizontal displacement values are mostly consistent within 25% with PPP displacement values, the displacement vector directions do not agree very well with the PPP results accompanied by several degrees biases.

We derived six fault slip distributions using the coseismic displacements obtained from the converged PPP solution, TPP-POPC solution, TPP-BOBC solution, variometric-POPC

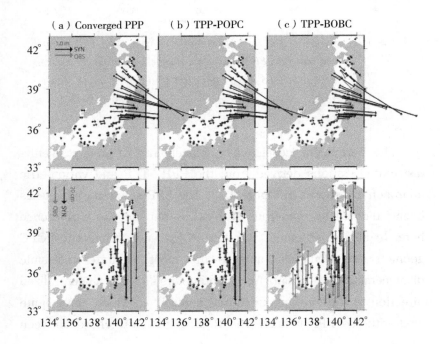

4.4 Single-Receiver Approaches for Real-Time GNSS Seismology

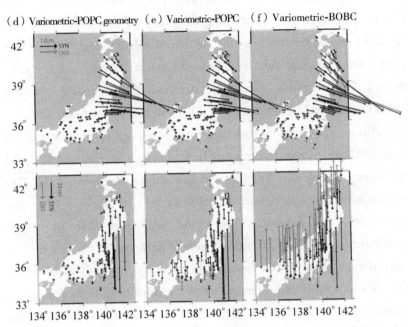

Figure 4.23 *The comparisons of the observed and synthetic coseismic displacements on horizontal components, and on vertical components, respectively. (a) Inversion with permanent displacements obtained from converged PPP solution; (b) Inversion with TPP-POPC solution; (c) Inversion with TPP-BOBC solution; (d) Inversion with variometric-POPC geometry solution; (e) Inversion with variometric-POPC solution; (f) Inversion with variometric-BOBC solution.*

geometry solution, variometric-POPC solution, and variometric-BOBC solution, respectively. The inversions are carried out using a FORTRAN code "SDM" based on the constrained least squares method (Wang et al., 2011). A priori conditions and physical constraints are chosen as the same as Wang et al. (2013). The total rupture area is assumed to be 650 km along the strike direction and 300 km along the dip direction, which is then divided into 1950 sub-faults with length and width of 10 km and 10 km, respectively. The dip angle linearly increases from 10° on the top (ocean bottom) to 20° at about 80 km depth. The rake angle (slip direction relative to the strike) is allowed to vary ±20° around 90°. Green's functions are calculated based on the CRUST2.0 model (Bassin et al., 2000) in the concerning area.

The comparisons of synthetic and observed displacements on horizontal and vertical components are shown in Figure 4.23, and the inverted fault slip distributions are shown in Figure 4.24. The best resolved slip model is assumed to be that derived from the converged PPP dataset. This model indicates that the peak coseismic slip of the earthquake nearly reached 23 m which is in agreement with previous results obtained by Wang et al. (2013) with onshore GPS data. The moment magnitude of the earthquake is estimated to be Mw 8.97, which is similar to the moment solution of about Mw 9.0, estimated by the USGS. Both the inversion results derived from TPP-POPC and variometric-POPC geometry datasets are quite consistent with the PPP slip model not only in the slip distribution pattern and moment magnitude, (Mw 8.96 for both), but also in the displacement fittings. For other three inversions of TPP-BOBC, variometric-POPC and variometric-BOBC (about Mw 8.90), there are obvious differences not only for the slip distribution patterns, but also for the displacement fittings. The peak slip values for the TPP-BOBC and variometric-POPC models are less than 21 m. Overall, the comparison of the six inversion results shows that the TPP-POPC and the variometric-POPC geometry solutions can derive reliable fault slip distribution, having consistent performance with the results inverted from converged PPP solution.

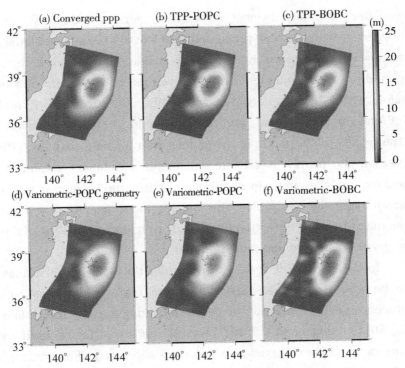

Figure 4.24 *Fault slip distributions for the 2011 Tohoku earthquake inverted from different permanent coseismic displacements obtained by different strategies: (a) converged PPP; (b) TPP-POPC; (c) TPP-BOBC; (d) variometric-POPC geometry; (e) variometric-POPC; (f) variometric-BOBC.*

4.5 Conclusions

A new approach for real-time GNSS seismology using a single receiver was presented. The performance of the proposed TPP approach is validated using 1 Hz GEONET data collected during the 2011, Mw 9.0 Tohoku-Oki earthquake.

When real-time precise orbit and clock corrections are

available, the displacement waveforms, derived from TPP, are consistent with the post-processed PPP waveforms at an accuracy of few centimeters during the entire shaking period, even for a period of twenty minutes. The TPP permanent coseismic offsets agree with PPP ones very well with RMS values of 3.0, 2.1, and 5.6 cms in north, east, and vertical components, respectively. The results of the fault slip inversions also indicate that the TPP method can provide a reliable estimation of moment magnitude and even of the fault slip distribution. If just the broadcast orbits and clocks are available, the displacement accuracy will be degraded to some extent and this leads to underestimations of the moment magnitude and fault slip values.

In the discussion of 4.2, we consider the GNSS observations to be free from cycle slips during the earthquake period. In practice, several methods of cycle-slip fixing (e.g., Zhang and Li, 2012; Geng et al., 2010; Li et al., 2013b) can be used to correct the phase observations when cycle slips occur. Furthermore, a joint processing of multi-GNSS (e.g., GPS, GLONASS, Galileo and BeiDou) data will significantly increase the number of available satellites and thus enhance the reliability of our approach.

In Section 4.4, we also compared the technical details of current single-receiver GNSS seismology approaches. Furthermore, several refinements are proposed to the variometric approach in order to eliminate the drift trend in the integrated coseismic displacements. We discussed the mathematical relationship among the PPP, TPP and refined variometric approaches and verified their equivalence based on two conditions: one is that all the error components in the TPP and variometric approaches are carefully considered following the PPP model; the other is that both TPP and variometric approaches use accurate known coordinates at the initial epoch (before the earthquake) to eliminate the geometry error.

We carefully analyzed the impact of error components such as satellite ephemeris, ionospheric delay, and geometry change on the displacements retrieved from the TPP and variometric approaches. The ionospheric delay has very significant impact on accumulative displacements and the drift values can reach up to several decimeters in horizontal components and about 1 meter in up components for 20 minutes integration interval. The satellite ephemeris, especially the satellite clock error, has critical influence on displacements which is depicted a complicated drift character with more fluctuations when broadcast orbit and clock is adopted. The geometry error also has a significant impact on accumulative displacements and the displacement error caused by geometry item can reach up few decimeters for 20 minutes interval.

We validated the performance of these single-receiver processing strategies (PPP, TPP and refined variometric approaches) using 1 Hz GPS data collected during the Tohoku-Oki earthquake (Mw 9.0, March 11, 2011) in Japan. After careful correction of all error components, the displacement waveforms derived from TPP and refined variometric approach are consistent with converged PPP at an accuracy of few centimeters. The results of the fault slip inversions also indicate that the TPP and the refined variometric approach can provide a reliable estimation of moment magnitudes and fault slip values as the converged PPP. From the above analysis and results on the single-receiver approaches for real-time GNSS seismology, we can conclude that the TPP, refined variometric approaches have equivalent mathematical model and can provide the same displacement precision with the converged PPP method. Moreover, these two approaches overcome the convergence problem of PPP, making them more suitable for seismological applications.

Chapter 5 Tightly-Integrated Processing of Raw GNSS and Accelerometer Data

5.1 Introduction

The problem with GPS displacement is that its noise level is much higher than that from most seismic sensors. In GPS displacements, this noise is basically white across the whole seismic frequency band. Seismic sensors measure acceleration with a very high precision and sampling rate and the seismic displacements can be obtained by double integration of the observed accelerometer signals. However, the acceleration is accompanied by unphysical drifts due to sensor rotation and tilt (Trifunac and Todorovska, 2001; Lee and Trifunac, 2009), hysteresis (Shakal and Petersen, 2001), and imprecision in the numerical integration process (Boore et al., 2002; Smyth and Wu, 2006). Its noise level, viewed in terms of displacement, will rise with decreasing frequency: at some frequency this noise level will exceed that of GPS. Therefore, GPS and seismic instruments can be mutually beneficial for seismological applications because weaknesses of one observation technique are offset by strengths in the other.

In order to take full use of the complementary of GPS and

seismic sensors, we propose an approach of integrating the accelerometer data into the precise point positioning (PPP, Zumberge et al., 1997) processing. Instead of combing the GPS-derived displacements with the accelerometer data (Emore et al., 2007; Bock et al., 2011), a tightly-integrated filter is developed to estimate seismic displacements from GPS phase and range and accelerometer observations. We apply the tightly-coupled integration to analyze collocated GPS and seismic data collected during the 2011 Tohoku-Oki (Japan) and the 2010 El Mayor-Cucapah (Mexico) earthquakes. Time and frequency domain analysis show that the integrated displacement and velocity waveforms are more accurate than GPS-only or seismic-only results. The integrated displacement waveform can capture both transient phenomena (waves) and permanent or static deformation. From the integrated results, we detect the P-wave arrival, locate the epicenter, and extract the permanent offsets for static slip inversion and magnitude estimation.

5.2 Overview of Combining GPS and Accelerometer Data

GPS relative kinematic positioning is usually adopted to estimate seismic displacements as double-differenced ambiguities can be fixed to integers for guaranteeing high accuracy (Larson et al., 2003; Blewitt et al., 2006; Crowell et al., 2012; Melgar et al., 2012; Ohta et al., 2012). In relative positioning, data from a network is analyzed simultaneously to estimate station positions with respect to at least one reference station which could also be displaced. PPP can provide "absolute" seismic displacements related to a global reference frame defined by the satellite orbits and clocks with a single GPS receiver (Kouba 2003; Wright et al., 2012). Especially, PPP integer ambiguity resolution, developed in recent years (Ge et al., 2008; Geng et al., 2012; Li and Zhang, 2012; Loyer et al., 2012), enables it to achieve comparable accuracy as relative positioning. Li et al. (2013a) demonstrated the performance of real-time PPP with ambiguity resolution using 5 Hz GPS data collected during El Mayor-Cucapah earthquake (Mw 7.2, 4 April, 2010) in Mexico.

Emore et al. (2007) estimated GPS displacements based on relative network analysis using the GPS analysis software GIPSY, developed by JPL (Jet Propulsion Laboratory) with orbits held fixed to precise IGS (International GNSS Service) final products. A constrained inversion technique was then used to combine GPS displacements and accelerometer data from the 2003 Mw 8.3 Tokachi-oki earthquake to estimate displacements and step function offsets in accelerometer records, after correcting for possible misorientation of the accelerometers.

A multirate Kalman filter was proposed by Smyth and Wu (2006) for fusing raw accelerometer with collocated GPS displacement data and was used for bridge monitoring (Kogan et al., 2008) and structural engineering applications (Chan et al., 2006). Bock et al. (2011) applied the multirate Kalman filter to estimate broadband displacements for the 2010 Mw 7.2 El Mayor-Cucapah earthquake by combining 1 Hz GPS displacements and 100 Hz data of collocated strong motion sensor in southern California. Hereby the 1 Hz GPS displacement was estimated using instantaneous GPS positioning in relative positioning mode (Bock et al., 2000). Geng et al. (2013) proposed a seismogeodetic approach and applied it to GPS and accelerometer observations of the 2012 Brawley seismic swarm. Melgar et al. (2013)

demonstrated the Kalman filter performance for the Tohoku-oki event and analyzed the spectral differences between GPS, Kalman and accelerometer data in detail. Wang et al. (2013) discussed the potential for an automated baseline correction scheme for accelerometer data that does not rely on GPS data.

In these combination procedures, the long-period stability of GPS derived positions is employed to constrain the seismic data. As is well-known, in kinematic positioning, precise dynamical information will give rather tight constraint on coordinates of adjacent epochs to strengthen the solution for more reliable ambiguity fixing and better displacement accuracy. The precise dynamical information of the movement provided by seismic sensors cannot be properly utilized to enhance GPS solutions if estimated coordinates are used. Therefore, integration on the observation level is required in order to have the advantages of both sensors and offset their weakness. In this study, the accelerometer data are integrated into the ambiguity-fixed PPP processing on the raw observation level.

5.3 The Tightly-Integrated Algorithm

Fixing ambiguities to integers can significantly improve the GPS positioning quality, especially for the east component (e.g., Blewitt, 1989; Dong and Bock, 1989). Due to the existence of uncalibrated phase delays (UPDs) originating at receiver and satellite (Blewitt 1989), for a long time only double-differenced ambiguities between satellites and receivers can be fixed. In the recent years, it was demonstrated that satellite UPDs could be estimated from a reference network and applied to other stations for fixing integer ambiguity in PPP mode (Ge et al., 2008;

Collins, 2008; Laurichesse et al., 2008; Li et al., 2013b). Thus, PPP with integer ambiguity fixing requires not only precise satellite orbit and high-rate satellite clock corrections but also UPDs product. There are several IGS real-time analysis centers providing UPDs product for PPP ambiguity fixing (Ge et al., 2012; Loyer et al., 2012). With the corrections of GPS satellite orbits, clocks and UPDs, the corresponding biases in the observations can be removed. The receiver-dependent UPD can be assimilated into receiver clock parameter. Hence, the linearized equations for raw carrier phase and pseudo-range observations then can be simplified as (Teunissen and Kleusberg, 1996),

$$l_j^s = -u^s \cdot \Delta r + m^s \cdot Z + t - I_j^s + \lambda_j N_j^s + \varepsilon_j^s \qquad (5.1)$$

$$p_j^s = -u^s \cdot \Delta r + m^s \cdot Z + t + I_j^s + e_j^s \qquad (5.2)$$

where, l_j^s, p_j^s denote "observed minus computed" phase and code observables from satellite s to receiver at frequency j; u^s is the unit direction vector from receiver to satellite; Δr denotes the vector of the receiver position increments; Z denotes tropospheric zenith wet delay; m^s is the wet part of global mapping function; t are the receiver clock errors; λ_j is the wavelength of the j frequency; I_j^s is ionospheric delay on the path at the j frequency; N_j^s is the integer phase ambiguity; e_j^s is the pseudo-range measurement noise; ε_j^s is measurement noise of carrier phase. Other error components such as the dry tropospheric delay, phase center offsets and variations, phase wind-up, relativistic effect and tide loading could be corrected with existing models (Kouba and Héroux, 2001).

Usually the ionosphere-free linear combination is employed in PPP to eliminate the effect of ionospheric delays. In order to suppress the measurement noise, instead of such linear combination we use in this contribution raw carrier-phase and pseudo-range observations at L1 and L2 frequencies (Schaffrin and Bock, 1988). The slant ionospheric delays are estimated as unknown parameters and a temporal constraint is introduced to strengthen the solution. Assuming that n satellites are observed by the receiver at the epoch k, the observational equations for all the satellites at this epoch can be expressed as,

$$Y_k = A_k \cdot X_k + \varepsilon_{Y_k}, \; \varepsilon_Y \sim N(0, Q_Y) \quad (5.3)$$

$$Y = \begin{pmatrix} (L_1^T, L_2^T)T \\ (P_1^T, P_2^T)T \end{pmatrix}, L_j = (l_j^1, \cdots, l_j^n)^T, P_j = (p_j^1, \cdots, p_j^n)^T \quad (5.4)$$

The design matrix and unknown parameters are:

$$A = \left(\begin{pmatrix} j_2 \\ j_2 \end{pmatrix} \otimes A', \begin{pmatrix} j_2 \\ j_2 \end{pmatrix} \otimes j_n, \begin{pmatrix} -\kappa \\ \kappa \end{pmatrix} \otimes J_n, \begin{pmatrix} \lambda \\ 0_2 \end{pmatrix} \otimes J_n \right) \quad (5.5)$$

$$A' = \begin{pmatrix} -u^1 & 0 & m^1 \\ \vdots & \vdots & \vdots \\ -u^n & 0 & m^n \end{pmatrix}, \; \kappa = \begin{pmatrix} 1 \\ \lambda_2^2/\lambda_1^2 \end{pmatrix}, \; \lambda = \begin{pmatrix} \lambda_1 & 0 \\ 0 & \lambda_2 \end{pmatrix} \quad (5.6)$$

$$X = (\Delta r^T \; \Delta \dot{r}^T \; Z \; t \; (I_1^s)^T (N_1^s)^T (N_2^s)^T)^T, \; (s = 1, \cdots, n), \quad (5.7)$$

where $\Delta \dot{r}$ denotes the vector of the receiver velocity; J_n is an identity matrix of n dimension; j_n denotes a column vector of n dimension in which all of the elements are unity; $\otimes$ is the Kronecker product; Q_Y is the variance-covariance matrix of ε_Y; κ is the coefficient of ionospheric delay.

The state equation can be described by:

$$X_k = \Phi_{k-1} \cdot X_{k-1} + \psi_{k-1} \cdot a_{k-1} + \varepsilon_{S\;k-1}, \; \varepsilon_S \sim N(0, Q_S) \quad (5.8)$$

$$\Phi = \begin{pmatrix} J_3 & \tau \cdot J_3 & & & & \\ & J_3 & & & & \\ & & 1 & & & \\ & & & 0 & & \\ & & & & J_n & \\ & & & & & J_{2n} \end{pmatrix}, \; \psi = \begin{pmatrix} \dfrac{\tau^2}{2} \cdot J_3 \\ \tau \cdot J_3 \\ 0_{1*3} \\ 0_{1*3} \\ 0_{n*3} \\ 0_{2n*3} \end{pmatrix} \quad (5.9)$$

$$Q_s = \begin{pmatrix} \frac{\tau^3}{3} \cdot q_a & \frac{\tau^2}{2} \cdot q_a & & & & & \\ \frac{\tau^2}{2} \cdot q_a & \tau \cdot q_a & & & & & \\ & & \tau \cdot q_z & & & & \\ & & & \tau \cdot q_t & & & \\ & & & & \tau \cdot q_i & & \\ & & & & & 0_{2n} \end{pmatrix}, \quad (5.10)$$

where, Φ is the system dynamics matrix; a is the system inputs vector (raw accelerometer observations from the seismic sensor); Ψ is the input matrix; τ is the accelerometer sampling interval; Q_s is the variance-covariance matrix of ε_S; q_a is the acceleration variance (1,000 times the pre-event noise of 60 s in this paper); q_z, q_t and q_i are the variances for the zenith wet delay (about 2-5 mm/$\sqrt{hour}$), receiver clock (is set to white noise with a very large value) and ionospheric delay (generally a few millimeters for the 5 Hz data sampling) respectively.

With the GPS observational equations of (5.3) and the state equations of (5.8), the real-time Kalman filter can be employed to estimate the unknown parameters,

$$\bar{X}_k = \Phi_{k-1} \cdot \hat{X}_{k-1} + \psi_{k-1} \cdot a_{k-1} \quad (5.11)$$

$$\bar{Q}_k = \Phi_{k-1} \cdot Q_{k-1} \cdot \Phi_{k-1}^T + Q_{S_{k-1}} \quad (5.12)$$

$$\hat{X}_k = \bar{X}_k + Q_k A_k^T Q_{Y_k}^{-1} \cdot (Y_k - A_k \cdot \bar{X}_k) \quad (5.13)$$

$$Q_k = (\bar{Q}^{-1}{}_k + A_k^T \cdot Q_{Y_k}^{-1} \cdot A_k)^{-1} \quad (5.14)$$

The time update of (5.11) and (5.12) is performed at every accelerometer sampling, while the measurement update of (5.13) and (5.14) is applied at every GPS epoch. The time and particularly the frequency domain performance of the filter can also be improved in post-processing with a smoother and in near real-time with a fixed lag smoother (Bock et al., 2011; Melgar et al., 2013).

The integer ambiguity resolution is attempted at every GPS epoch, L1 and L2 ambiguities are fixed simultaneously using the LAMBDA method by Teunissen (1995). With the predicted ionospheric delays from previous ambiguity-fixed epochs, reliable ambiguity resolution is achievable within few seconds for re-convergence (e.g., Geng et al., 2010; Zhang and Li, 2012; Li et al., 2013b), although a convergence period of about 20 min for ambiguity fixing is still required.

Another critical issue is the validation of the fixed integer ambiguities. There are several approaches to assess the resolved integer ambiguities, such as R-ratio, W-ratio as well as the Integer Aperture-based R-ratio, and W-ratio methods (Li and Wang, 2012). In this study, the well-known R-ratio test was used to validate the ambiguity resolution. The R-ratio is defined as the proportion of the second minimum and the minimum quadratic distances between the integer and the real-valued ambiguities. It is used to discriminate between the second set of optimum integer candidates and the optimum one usually with a critical criterion of three (Han, 1997).

5.4 Results

The 2011 Mw 9.0 Tohoku-Oki earthquake (11 March, 2011, 05:46:24 UTC) in Japan and the 2010 Mw 7.2 El Mayor-Cucapah earthquake (4 April, 2010, 22:40:42 UTC) in Mexico were well recorded not only by strong motion stations but also by high-rate GPS receivers. They are good examples to evaluate the performance of integrated displacements for which abundant high-rate GPS and strong motion records are available.

We firstly processed 1 Hz data of about 90 globally distributed

Chapter 5 Tightly-Integrated Processing of Raw GNSS and Accelerometer Data

real-time IGS stations using the EPOS-RT software of GFZ in simulated real-time mode for providing GPS orbits, clocks and UPD corrections at 5 s sampling interval. Based on these corrections, we replay the GPS and strong motion data collected at about thirty collocated stations during the Tohoku-Oki and El Mayor-Cucapah earthquakes. As PPP can be performed with a single GPS receiver, the integrated displacements are estimated on a pair-by-pair basis for each collocated GPS and strong motion pair.

5.4.1 Comparison of GPS, Seismic and Integrated Waveforms

The 2010 Mw 7.2 El Mayor-Cucapah earthquake (4 April, 2010, 22:40:42 UTC) in northern Baja California, Mexico, provides us with a real event to evaluate the performance of the proposed tightly-integrated approach. GPS data is collected from the California Real-Time Network (CRTN, Genrich and Bock, 2006) and Plate Boundary Observatory (PBO, Jackson, 2003). 200 Hz accelerometer data is collected from strong motion stations of the Southern California Seismic Network (SCSN) operated by the USGS (U.S. Geological Survey) and Caltech. Table 5.1 summarizes the station names, locations, distances to epicenter and separations for four collocated stations analyzed.

Displacement waveforms are estimated for each collocated pair of GPS and strong motion sensors using the presented tightly-integrated filter, in a simulated real-time mode. We compare the integrated displacements with GPS-only displacements derived from real-time ambiguity-fixed PPP and seismic-only displacements obtained through double integration of seismic accelerations. The seismic-only displacements in this study are provided by California Geological Survey (CGS/CSMIP, http://strongmotioncenter.org/).

Table 5.1 Collocated high-rate GPS and strong motion (SM) stations

Station (GPS/SM)	Latitude (°N)	Longitude (°E)	Dist to Epic (km)	Separation (km)
P500	32.690	−115.300	48.4	0.56
NP5054	32.693	−115.338		
P496	32.751	−115.596	61.8	0.07
NP5058	32.752	−115.595		
P744	32.829	−115.508	66.4	0.14
NP5028	32.829	−115.505		
P499	32.980	−115.488	83.9	1.80
NP5060	32.991	−115.513		

The baseline offsets are already corrected by applying a high-pass filter. The GPS station P496, which is located about 60 km from the epicenter, is collocated with SCSN seismic station 5058 (about 140 m distance). Comparison of the GPS-only, seismic-only and tightly integrated displacements in all three components for this pair (5058/P496) is exemplarily shown in Figure 5.1 by black, blue and red line, respectively.

In Figure 5.1a, we show the entire period of seismic shaking in the north component. The GPS-only and seismic-only displacements show a high degree of similarity of the dynamic component. The standard deviation (STD) values of the differences between GPS-only and seismic-only displacements are found to be 1.1, 1.0 and 2.1 cms respectively in north, east and vertical components. The obvious difference is that a permanent coseismic offset of 0.2 m is visible in the GPS-only displacements. Tilt and rotation of the seismic instrument result in distortions and

Chapter 5 Tightly-Integrated Processing of Raw GNSS and Accelerometer Data

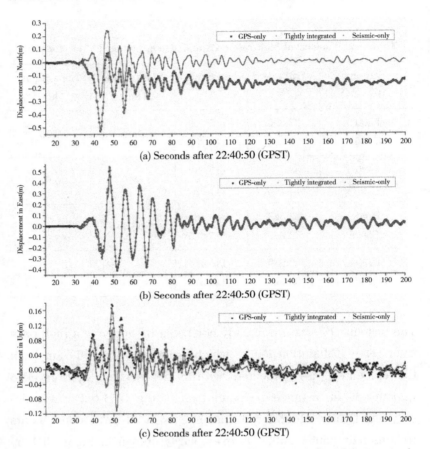

Figure 5.1 *Comparison of GPS-only, seismic-only and tightly-integrated displacements on the collocated 5058 (seismic) and P496 (GPS) stations during the El Mayor-Cucapah earthquake on 4 April 2010. The sub-figures (a), (b), (c) show the entire period of seismic shaking in the north, east and up components respectively. The 5 Hz GPS-only, 200 Hz seismic-only, and 200 Hz tightly-integrated displacements are respectively shown by the black, blue and red lines.*

baseline offsets. Although these effects are largely removed by high-pass filter, low-frequency information is lost, including the loss of permanent coseismic offset in the seismic-only displacements (Allen and Ziv, 2011). In the tightly integrated displacements, the 0.2 m permanent offset in the north component is clearly seen as the seismic data is constrained by the long-period stability of GPS measurements and the baseline-shift problem in seismic data can be overcome. The 5 Hz GPS-only displacements are with lower sampling rate and higher noise compared to the 200 Hz seismic-only displacements. The root mean square (RMS) values of GPS-only solution (10 min pre-event displacement series) are found to be 1.1, 1.1 and 3.0 cms respectively in north, east and vertical components. The vertical component (Figure 5.1c) is the noisiest as expected, due to the satellite constellation configuration and the high correlation between zenith tropospheric delays and the vertical component. With the aid of the accelerometer data, the tightly-integrated filter is capable of producing a more precise waveform. The small-amplitude seismic signal can be detected from the tightly integrated solution (e.g. from 20 s to 35 s in Figure 5.1c and Figure 5.2c) in spite of the diminished precision of the GPS vertical component. This is a significant improvement compared to the GPS-only solution where earthquake signal is detected only for strong events with significant shaking.

For clarity, the blowup of the displacement series in all three components for its first 45 s is shown in Figure 5.2. We can see that the tightly-integrated displacements are in good agreement with GPS-only solution in terms of peak displacements and long-period stability. Meanwhile, the displacement precision is also improved by precise dynamical information provided by seismic sensors. The small-amplitude details of the movement (e.g., small shakes around 47 s in Figures 5.2a, 5.2b, and 5.2c), which are often covered by measurement noise in GPS-only solution, can be clearly observed from the tightly-integrated waveform.

In Figure 5.3, we show the differences between tightly-integrated

Chapter 5 Tightly-Integrated Processing of Raw GNSS and Accelerometer Data

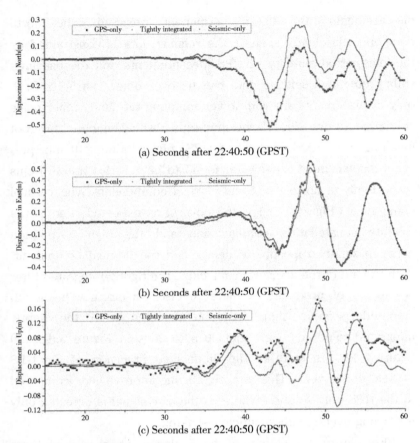

Figure 5.2 *The blowup of the first 45 s of the coseismic displacements in all three components on the collocated 5058 (seismic) and P496 (GPS) stations during the El Mayor-Cucapah earthquake on 4 April, 2010. The north, east and up components are shown in the sub-figures a, b, and c, respectively.*

filter and GPS-only/seismic-only displacements for the 5058/P496 pair. The results for the north, east and up components are respectively shown in Figures 5.3a, 5.3b and 5.3c. One can see that the differences between GPS-only and filter displacements show a high frequency noise due to the diminished precision of

5.4 Results

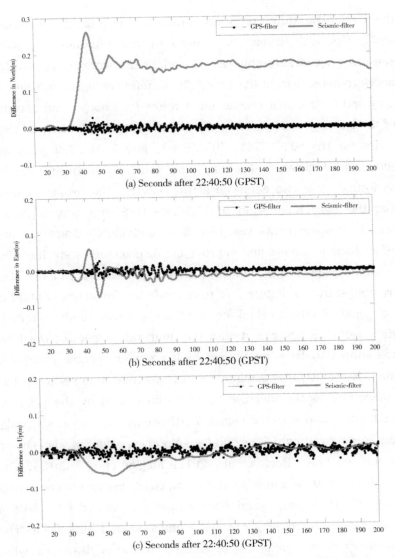

Figure 5.3 *Difference between the displacement series from tightly-integrated filter and each of the two inputs (GPS and seismic displacements) for 5058/P496 pair. The differences between GPS-only and filter displacements are shown by the black line, while the differences between seismic-only and filter displacements are shown by the red line. The differences in north, east and up components are respectively shown in the sub-figures a, b, and c.*

the GPS. The differences are more scatter during the strong shaking period. It may be caused by the separation between sensors (not strictly collocated) and/or an overweighting of the accelerometer data in the filter. The differences between seismic-only and filter displacements show a low-frequency trend because of the baseline-shift problem of seismic data. The differenced time series for the 5028/P744, 5054/P500, and 5060/P499 pairs are shown in Figures 5.4, 5.5, and 5.6, respectively. Similar performance is also achieved at these pairs, the results confirm that the tight integration of high-rate GPS and very high-rate seismic measurements can take their individual advantages and offset their weakness and improve the displacements significantly.

Power spectral densities for filter displacements at P496/5058 are also shown in Figure 5.7a to quantify the frequency content of the signal. Similar to Bock et al. (2011), a saw tooth pattern in the waveforms associated with the multi-rate aspect of the filter was shown to have an impact in the power spectra (the blue line). However, it is a minor problem for real-time seismological applications as the increase in noise introduced by the peaks is small compared with the signal. Furthermore, the spurious peaks can be removed by a 5-second lag smoother, which is also shown by the red line (Bock et al., 2011; Meglar et al., 2013). The power spectral densities of the 3 kinds of displacements (5 Hz GPS, 200 Hz seismic, and 200 Hz tightly-integrated filter with a 5-second lag smoother) are also compared in Figure 5.7b. The frequency domain analysis of these waveforms describes what frequency bands each data type is reliable in: GPS performs better at the lower frequencies and accelerometer is better at the higher frequencies. The power spectral densities of filter displacements follow the GPS-only spectrum at the low frequencies and the accelerometer-only spectrum at the high frequencies. From the

5.4 Results

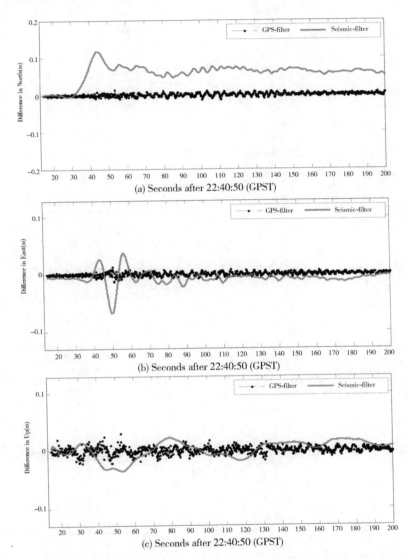

Figure 5.4 *Difference between the displacement series from tightly-integrated filter and each of the two inputs (GPS and seismic displacements) for 5028/ P744 pair. The differences between GPS-only and filter displacements are shown by the black line, while the differences between seismic-only and filter displacements are shown by the red line. The differences in north, east and up components are respectively shown in the sub-figures a, b, and c.*

Chapter 5 Tightly-Integrated Processing of Raw GNSS and Accelerometer Data

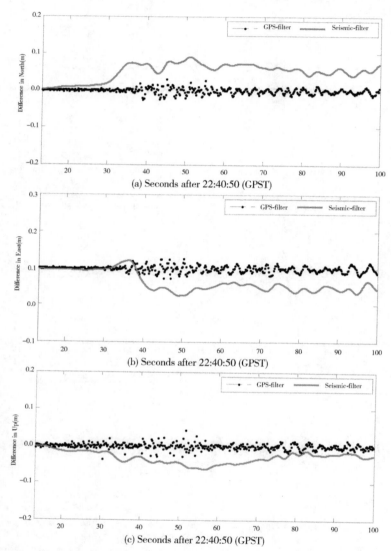

Figure 5.5 *Difference between the displacement series from tightly-integrated filter and each of the two inputs (GPS and seismic displacements) for 5054/ P500 pair. The differences between GPS-only and filter displacements are shown by the black line, while the differences between seismic-only and filter displacements are shown by the red line. The differences in north, east and up components are respectively shown in the sub-figures a, b, and c.*

5.4 Results

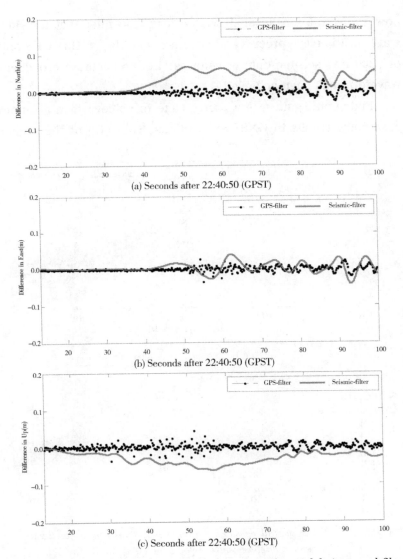

Figure 5.6 *Difference between the displacement series from tightly-integrated filter and each of the two inputs (GPS and seismic displacements) for 5060/ P499 pair. The differences between GPS-only and filter displacements are shown by the black line, while the differences between seismic-only and filter displacements are shown by the red line. The differences in north, east and up components are respectively shown in the sub-figures a, b, and c.*

power spectral density analysis, we can also infer that the filter waveform is more precise and accurate than the (5 Hz) GPS-only or (200 Hz) seismic-only waveforms, i.e. an accurate broadband waveform has been achieved.

Fixing ambiguities is a prerequisite to achieve high accuracy positioning results in GNSS applications. The ratio of the second

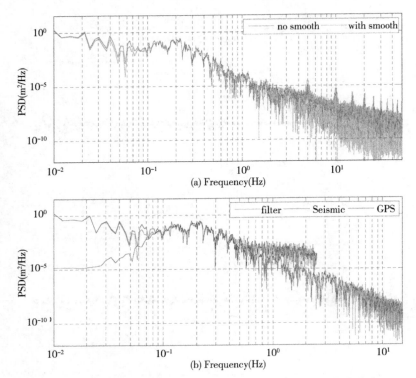

Figure 5.7 *Power spectral densities. (a) Power spectral density for tightly-integrated displacement waveforms at P496/5058. The blue line denotes the results without a smoother, while the red line denotes the results with a 5-second lag smoother. (b) Power spectral density for 5 Hz GPS displacements at P496 (the black line), 200 Hz high-pass filtered seismic displacements at 5058 (the blue line), and 200 Hz tightly-integrated displacements (with a 5-second lag smoother) at P496/5058 (the red line).*

minimum to the minimum quadratic form of residuals (R-ratio) is used here to decide the correctness and confidence level of integer ambiguity candidate. The ratio value can be considered as an index to denote the reliability of ambiguity resolution. Thus, larger ratio values denote more reliable ambiguity resolution. The ratio values of tightly-integrated and GPS-only solution for 5058/P496 and 5028/P744 pairs are respectively shown in Figures 5.8a and 5.8b. The ratio values of tightly-integrated solution are shown by the red line, while the ratio values of GPS-only solution are shown by the black line. As shown in Figure 5.8, the ratio values of GPS-only solution are generally rather small and below 5

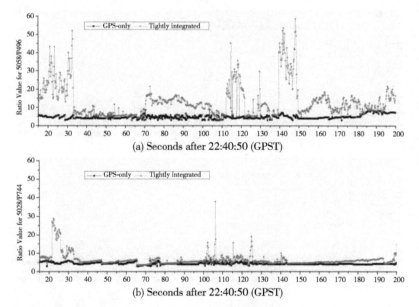

Figure 5.8 Comparison of the ratio values for tightly-integrated and GPS-only solutions. (a) for the collocated 5058 (seismic) and P496 (GPS) pair; (b) for the collocated 5028 (seismic) and P744 (GPS) pair. The ratio values of tightly-integrated and GPS-only solutions are shown by the red and black line, respectively.

usually. With the aid of the accelerometer data, the ratio values are increased remarkably compared to that of GPS-only. The averaged ratio is increased from 4.5 and 3.8 of GPS-only for 5058/ P496 and 5028/P744 pairs to 11.6 and 6.4 of the tightly-integrated solution, respectively. The results indicate that the proposed algorithm can significantly improve the ability of resolving integer-cycle phase ambiguities, which is very critical for promoting the contribution of GPS phase observations. The comparison of ratio values for 5054/P500 and 5060/P499 pairs are also respectively shown in Figures 5.9a and 5.9b. The average values of GPS-only ratio values for these two pairs are improved from about 3.6 and

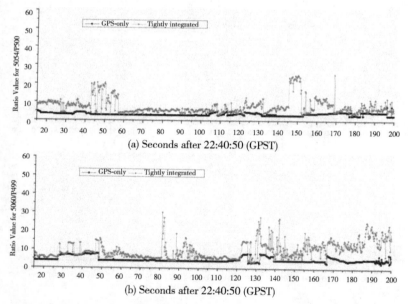

Figure 5.9 *Comparison of the ratio values for tightly-integrated and GPS-only solutions. (a) for the collocated 5054 (seismic) and P500 (GPS) pair; (b) for the collocated 5060 (seismic) and P499 (GPS) pair. The ratio values of tightly-integrated and GPS-only solutions are shown by the red and black line respectively.*

4.8 to 8.0 and 10.1, respectively.

For the 2011 Tohoku-Oki earthquake, the 1 Hz GPS data is collected at the GEONET (the GPS Earth Observation Network System) stations operated by the Geospatial Information Authority (GSI) of Japan. 100 Hz accelerometer data is collected from strong motion stations of the K-Net and Hi-Net. We compare the integrated displacements with seismic-only waveforms obtained from double integration of raw acceleration data. The results of two collocated pairs AKT006/0183 and NGN017/0986 are shown in Figure 5.10 as an example. The left sub-figures show the entire period of the seismic shaking in north/east/up components at AKT006/0183 and the right ones show the seismic shaking at NGN017/0986 in the same three components. The GPS station 0183 ($40.2154°$ N, $140.7873°$ E), which is located 251 km from the epicenter of Tohoku-Oki earthquake, is collocated with K-Net seismic station AKT006 (about 20 meters away from GPS station), and the other pair NGN017 and 0986 station within 5 km distance, where the distance to the epicenter is about 480 km.

The uncorrected seismic displacements are traditionally observed from zero-order corrected with only consideration removing the pre-event mean bias. Although the dynamic motions can be determined, a linear or parabolic drift is apparent in the latter part of each displacement time series, and the permanent coseismic offset is lost in a seismic-only solution. The corrected seismic displacements are derived from the baseline-corrected strong motion recordings which are processed using the automatic empirical baseline correction scheme proposed by Wang (2000). Although the corrected seismic displacements have a high degree of similarity of the dynamic component with the integrated results, they still maintain several decimeter differences in permanent coseismic offsets due to the effect of the residual baseline bias error. From the integrated displacement waveforms, there are obvious permanent coseismic offsets which are about 0.47 m, 0.51 m, and 0.03 m in the north, east, and up components at station AKT006/0183, while the permanent offsets of station NGN017/0986 are relatively small, about 0.04 m in the north, 0.12 m in the east, and 0.01 m in the up components.

In Figure 5.11 and Figure 5.12, we compare the tightly-integrated displacements (the red line) and GPS-only displacements (the black cross symbols). The results of the AKT006/0183 and NGN017/0986 pairs are respectively shown in the left and right side of Figure 5.11, and the similar results of the 5058/P496 and 5028/P744 pairs are also shown in Figure 5.12. The GPS station P496, which is located about 60 km from the epicenter of 2010 El Mayor-Cucapah earthquake, is collocated with SCSN seismic station 5058 (about 70 m separation). The other pair P744 and 5028 station are within 140 m of each other, and the distance from them to the epicenter is about 65 km. All sub-figures from top to bottom depict the entire period of seismic shaking in north, east and up components. We can see that the integrated displacements are in good agreement with GPS-only solution in terms of peak displacements, permanent offsets and long-period stability. However, it is clearly shown that the GPS-only displacements are with lower sampling rate and higher noise compared to the integrated displacements (Details can also be seen in Figure 5.14 and Figure 4.15). The root mean square (RMS) values of GPS-only solution (10 mins pre-event displacement series) are 1.1, 1.1 and 3.0 cms respectively in north, east and vertical components. The precision of integrated displacement is significantly improved by precise dynamical information provided by seismic sensors.

5.4 Results

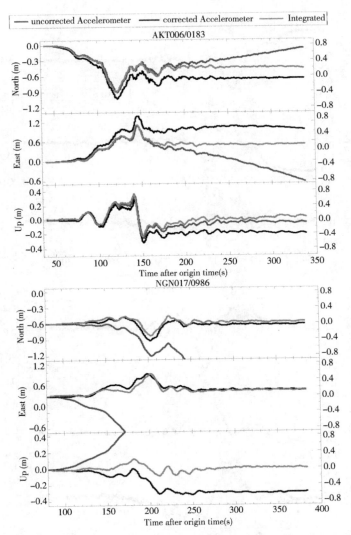

Figure 5.10 *Comparison of uncorrected seismic-only and corrected seismic-only and integrated displacements on the collocated AKT006 (seismic) and 0183 (GPS) pair and NGN017 (seismic) and 0986 (GPS) pair during the Tohoku-Oki earthquake on 11 March, 2011. All sub-figures show the entire period of seismic shaking. The 100 Hz integrated displacement is shown by the red line. The 100 Hz seismic-only displacements without baseline correction and with baseline correction are respectively shown by the blue line and the black line, respectively.*

Chapter 5 Tightly-Integrated Processing of Raw GNSS and Accelerometer Data

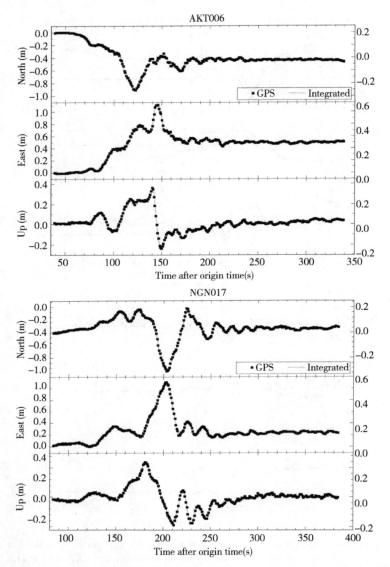

Figure 5.11 *Comparison of GPS-only and tightly-integrated displacements on the collocated AKT006 (seismic) and 0183 (GPS) pair and NGN017 (seismic) and 0986 (GPS) pair during the Tohoku-oki earthquake on 11 March 2011. The sub-figures show from top to bottom the entire period of seismic shaking in north, east and up components respectively. The 1 Hz GPS-only and 100 Hz tightly-integrated displacements are shown respectively by the black crosses and the red line.*

5.4 Results

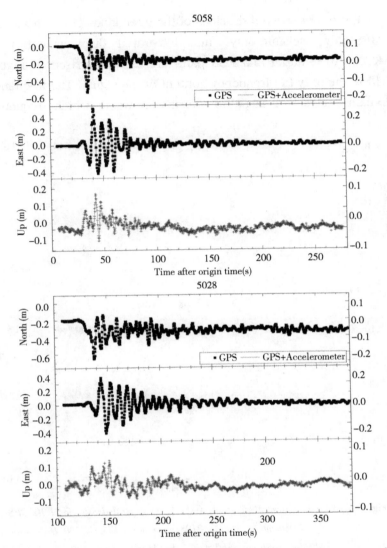

Figure 5.12 *Comparison of GPS-only and tightly-integrated displacements on the collocated 5058 (seismic) and P496 (GPS) pair and 5028 (seismic) and P744 (GPS) pair during the El Mayor-Cucapah earthquake on 4 April 2010. The sub-figures from top to bottom show the entire period of seismic shaking in north, east and up components respectively. The 5 Hz GPS-only and 200 Hz tightly-integrated displacements are shown respectively by the black cross symbols and red lines.*

Chapter 5 Tightly-Integrated Processing of Raw GNSS and Accelerometer Data

The power spectral densities of the three kinds of displacements (GPS-only, seismic-only, and integrated displacements) at AKT006/0183 and P744/5028 pairs are also compared in Figure 5.13 to quantify the frequency content of the signal. The frequency domain analysis of these waveforms shows in which frequency

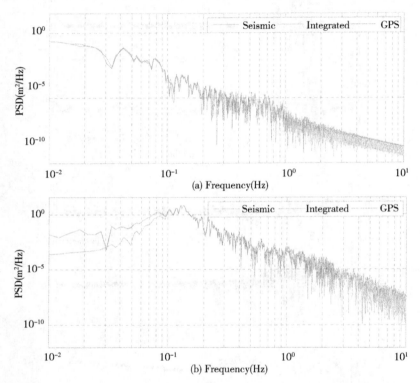

Figure 5.13 Power spectral densities. (a) Power spectral density for 1 Hz GPS displacements at station 0183 (the black line), 100 Hz seismic displacements at AKT006 (the blue), and 100 Hz tightly-integrated displacement waveforms at AKT006/0183 (the red line). (b) Power spectral density for 5 Hz GPS displacements at P744 (the black line), 200 Hz seismic displacements at 5028 (the blue line), and 200 Hz tightly-integrated displacements at P744/5028 (the red line).

bands each data type is reliable. GPS performs better at lower frequencies and seismic sensor is better at higher frequencies. We can see that the power spectral densities of integrated displacements follow the GPS-only spectrum at the low frequencies and the seismic-only spectrum at the high frequencies. From the power spectral density analysis, we can also infer that the integrated waveform is more precise and accurate than the GPS-only or seismic-only waveforms. An accurate broadband waveform, which has the advantages of both sensors, has been achieved.

5.4.2 Detection of P-Wave Arrival

Earthquake monitoring and early warning systems not only depend on the accurate estimation of permanent displacements, but also rely on the capability of the sense of P-wave arrival which is employed to predict the arrival and intensity of destructive S and surface waves. Figure 5.11 and Figure 5.12 have shown that the integrated results could get accurate permanent offsets. The following sections mainly focus on another issue. The enlarged view of the first 20 seconds of the integrated and GPS-only results for station 5028/P744 is shown in Figure 5.14, and the similar enlarged view for station AKT006/0183 is shown in Figure 5.15. From coseismic displacement and velocity waveforms, we can observe that the GPS-only solution is noisy and has a precision limited to several millimeters in displacement and few centimeters per second in velocity. The vertical component is much noisier as expected, due to the satellite constellation configuration and the high correlation between zenith tropospheric delay and the height component. The precision of vertical displacement is of the order of few centimeters, and vertical velocity precision is around several centimeters per second, which is not enough to detect P-wave accurately. With the aid of the seismic data, the tightly-integrated filter is capable of producing a precise integrated displacement and velocity waveform, especially in the up component. We can observe the small-amplitude P-wave in the recordings which are often covered by measurement noise in GPS-only solution, can be clearly observed from the integrated waveform, and detect P-wave arrival from the integrated solution in spite of the diminished precision of the GPS vertical component. This is a significant improvement over the GPS-only solution where earthquake signal is detected only after the S-wave arrival, which is generally a few seconds later than the P-wave arrival for near-field stations.

The bottom sub-figures in Figures 5.14 and 5.15 are STA/LTA ratio values based on tightly-integrated results for north/east/up components, which are used to pick up the earthquake P-wave arrival. The short-term average (STA) through long-term average (LTA) picker is the most broadly used automatic algorithm in seismology (Allen, 1978). It continuously calculates the average values of the absolute amplitude of a seismic signal in two consecutive moving-time windows. The short time window (STA) is sensitive to seismic events while the long time window (LTA) provides information about the temporal amplitude of seismic noise at the site (Trnkoczy et al., 2002). When the ratio of both exceeds a pre-set threshold means the arrival of P-wave. The STA/LTA picker parameter settings are always a tradeoff between several seismological and instrumental considerations. For these two earthquake events in this paper, the STA window duration is 0.2 seconds, the LTA window duration is 2 seconds, and the pre-set threshold is set to 10. We can clearly identify P-wave arrivals in the STA/LTA ratio time series. It is noted that the P-wave appears

5.4 Results

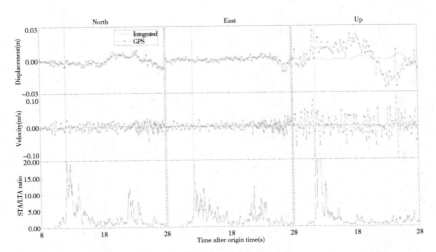

Figure 5.14 *An enlarged view of the first 20 s of the coseismic displacements and velocities in all three components on the collocated 5028 (seismic) and P744 (GPS) pair during the El Mayor-Cucapah earthquake. The 5 Hz GPS-only and 200 Hz tightly-integrated displacements and velocities are respectively shown by the blue dotted lines and red lines. The bottom sub-figures are STA/LTA ratio results based on tightly-integrated results, which show the first arrival time of seismic wave. The sub-figures show, from left to right, the north, east and up components.*

in vertical component first and in the horizontal components a few milliseconds later. The detected earthquake P-wave arrival time of station AKT006/0183 is 41.41s compared with the USGS reference value 41.55s calculated by TauP Toolkit (Crotwell et al., 1999), and the P-wave arrival time of station 5028/P744 is 11.49s compared with the reference value 11.58s. It is demonstrated that the integrated results could be used to pick up an accurate P-wave arrival time. However, it is difficult for the GPS-only solution to be accurately identified P-waves because of the significantly less

149

Chapter 5 Tightly-Integrated Processing of Raw GNSS and Accelerometer Data

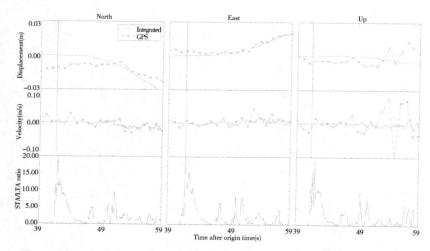

Figure 5.15 *An enlarged view of the first 20 s of the coseismic displacements and velocities in all three components on the collocated AKT006 (seismic) and 0183 (GPS) stations during the Tohoku-Oki earthquake on 11 March 2011. The 1 Hz GPS-only and 100 Hz tightly-integrated displacements and velocities are respectively shown by the blue dotted lines and red lines. The bottom sub-figures are STA/LTA ratio results based on tightly-integrated results, which show the first arrival time of seismic wave. The sub-figures show, from left to right, the north, east and up components.*

precision. Thus, the integrated result improves on both seismic-only and GPS-only methods, by providing the full spectrum of seismic motions from the detection of P-wave arrivals to the estimation of permanent offsets.

When P-wave is detected at four or more near-field GPS/strong motion pairs, the epicenter, the velocity of earthquake wave and the origin time can be determined by using a least squares method as follows,

$$\begin{cases} \left(\dfrac{x_1 - x_0}{d_1^0} - \dfrac{x_2 - x_0}{d_2^0}\right) \cdot dx + \left(\dfrac{y_1 - y_0}{d_1^0} - \dfrac{y_2 - y_0}{d_2^0}\right) \\ \quad \cdot dy + (t_1 - t_2) \cdot v - (d_1^0 - d_2^0) = 0 \\ \left(\dfrac{x_1 - x_0}{d_1^0} - \dfrac{x_3 - x_0}{d_3^0}\right) \cdot dx + \left(\dfrac{y_1 - y_0}{d_1^0} - \dfrac{y_3 - y_0}{d_3^0}\right) \\ \quad \cdot dy + (t_1 - t_3) \cdot v - (d_1^0 - d_3^0) = 0 \\ \quad \vdots \\ \left(\dfrac{x_1 - x_0}{d_1^0} - \dfrac{x_n - x_0}{d_n^0}\right) \cdot dx + \left(\dfrac{y_1 - y_0}{d_1^0} - \dfrac{y_n - y_0}{d_n^0}\right) \\ \quad \cdot dy + (t_1 - t_n) \cdot v - (d_1^0 - d_n^0) = 0 \\ t_0 = \dfrac{\sum_{i=1}^{n} \left(t_i - \dfrac{d_i}{v}\right)}{n} \end{cases} \quad (5.15)$$

Where, x_0, y_0 denote the approximate coordinates of the epicenter; $x_i, y_i (i = 1, \cdots n)$ denote the coordinates of the i the station; d_i^0 denotes the distance from the i th station to the approximate coordinates of epicenter; dx, dy denote the increments of epicenter; v denotes velocity of earthquake wave; t_i denotes the arrival time of earthquake wave at the i th station; d_i denotes the distance from the i th station to the epicenter; t_0 denotes the origin time. Several iterations are required to avoid the linearization error.

In order to test this technique, the five GPS/strong motion pairs where P-wave is detected earliest during the El Mayor-Cucapah earthquake are used. The detected earthquake P-wave arrival time is 0.09s, 0.15s, 0.11s, 0.10s, and 0.13s later than the USGS reference values of P-wave arrival time at the five pairs. The epicenter estimate is roughly 2.5 km away from the U.S. Geological Survey (USGS) epicenter estimate. The origin time estimate is 0.12s later than the USGS reference value of 22:40:57

(GPS time). The accurate detection of P-wave arrival is critical for earthquake early warning, as it allows for prediction of the arrival of the destructive S-wave. The P-wave-based earthquake parameters such as epicenter and origin time can be released before the S-wave arrival.

5.4.3 Extraction of Permanent Offset and Fault Slip Inversion

In addition to P-wave arrival time, the important information, provided by the integrated position series, is the permanent offset. We use the real-time algorithm proposed by Allen and Ziv (2011) to remove dynamic oscillations and extract these offsets. The permanent offsets derived from integrated solution (about 1 minute after the arrival of the earthquake wave) are compared with the ones from the post-processed daily solution in Figure 5.16. The RMS of the differences between them is about 3.7 mm.

We derived the spatial distribution of the fault slip using the coseismic displacements obtained from both the real-time tightly-integrated solution and the post-processed daily solution. In the same way as done by Li et al., (2014), the fault geometric parameters (strike 312°/dip 88°) are adopted from the Global Centroid Moment Tensor (GCMT) solution of the earthquake. The rake angle (slip direction relative to the strike) is allowed to vary ±20° around the GCMT solution of 186°. The fault size is given to be 130 km along the strike and 20 km along the dip, which is then divided into 26×4 = 104 sub-faults. In the inversion, the data is weighted twice as much for the two horizontal components as for the vertical component.

The inversion results are shown in Figure 5.17. The two inversions result in scalar seismic moments of 7.27×1019 Nm and 7.18×1019 Nm respectively, equivalent to moment magnitude of

5.4 Results

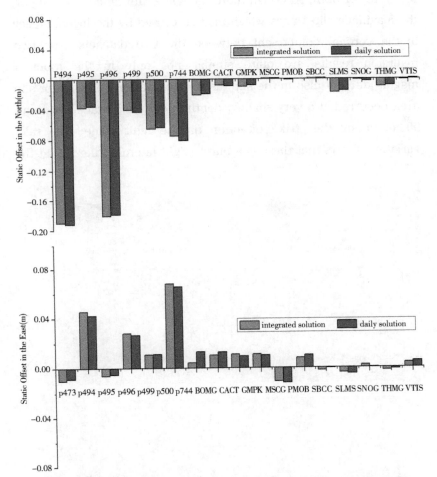

Figure 5.16   *Comparison of the permanent (static) offsets from the tightly-integrated solution and the post-processed daily solution. The blue rectangle shows static offsets derived from the static PPP solution with daily observations (the difference between daily solutions of the day before the earthquake and the day after the earthquake). The red rectangles show the static offsets derived from the real-time tightly-integrated solution.*

153

Mw 7.18 for both. Although there are some differences existing on the maximum slip values which may be caused by the inconsistency in the vertical component between the two datasets, the two inversion results are quite similar not only in the moment magnitude, but also in the slip distribution pattern. The major slip area occurred at a very shallow depth (near the surface) at about 90 km along the strike direction on the fault plane. The rake variation shows that there is a purely right lateral strike slip at the

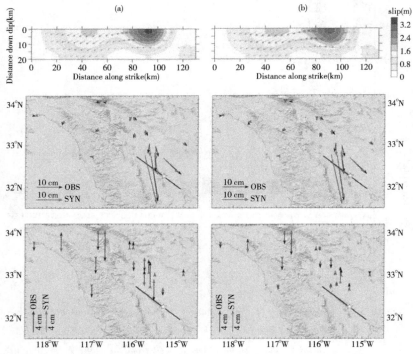

Figure 5.17 *Fault slip inversion. (a) Inversion with permanent coseismic displacements obtained from real-time tightly-integrated solution; (b) Inversion with post-processed daily solution. From top to bottom are the inverted fault slip distributions, comparisons between the observed and the synthetic displacements on the horizontal components, and on the vertical components, respectively.*

northwest of the fault, and a minor normal fault component occurs at the south east of the fault. Considering the hypocentral location, we can confirm that this earthquake is an asymmetric bilateral rupture event: the rupture mainly propagates northwestward from the hypocenter during the source process. Overall, the comparison of the two inversion results shows that the integrated solution can provide a reliable estimation of earthquake magnitude and even of the fault slip distribution in real time.

5.5 Conclusions

We presented an approach for tightly combining GPS and seismic sensor data where the accelerometer data are integrated into the ambiguity-fixed PPP processing on the observation level. The performance of the proposed tightly-integrated approach was validated using the collocated high-rate GPS and strong motion data collected during the 2010, Mw 7.2 El Mayor-Cucapah earthquake and the 2011 Tohoku-Oki (Japan) earthquake. For tightly-integrated displacements, the peak displacements and long-period stability are in agreement with GPS-only solution. As a typical example, the permanent coseismic offset which is usually underestimated in the seismic-only solution can be now obtained exactly in the integrated solution. Some small-amplitude seismic details, which are not detectable in the GPS-only approach, can be detected from the tightly integrated displacements. The integrated waveform takes the advantages of both sensors and is more precise and accurate than the GPS-only or seismic-only waveforms. A power spectral density analysis also demonstrates that an accurate broadband displacement waveform can be derived from the tightly-integrated filter. The power spectral densities of integrated displacements follow the GPS-only spectrum at the low frequencies and the seismic-only spectrum at the high frequencies. Furthermore, the ambiguity-fixing ratio values of the tightly-integrated solutions are significantly improved from about 4 of the GPS-only solutions to 10 on average.

The integrated displacements can provide the full spectrum of the seismic motion allowing the detection of P-wave arrivals and the estimation of permanent offsets. Small-scale features including P waves are visible in the integrated displacement and velocity waveforms. The P-wave arrival can be picked up accurately and used for reliable determination of epicenter and origin time. Permanent offsets can also be extracted with high accuracy and used for reliable fault slip inversion and magnitude estimation. These earthquake parameters are critical for earthquake/tsunami monitoring and early warning systems.

Chapter 6 Conclusions and Outlook

The main contributions of this book can be summarized as follows:

The PPP technique can provide "absolute" coseismic displacements with respect to a global reference frame with a stand-alone GPS receiver. In this paper, we present an approach of using real-time ambiguity-fixed PPP for earthquake early warning. The 2010 El Mayor-Cucapah earthquake and the 2011 Tohoku-Oki earthquake were used to evaluate our approach for possible use in EEW. Comparisons of the displacements, estimated from PPP and accelerometers, displayed a high agreement within few centimeters. Integer ambiguity fixing can improve the accuracy of real-time PPP displacements significantly, especially in the east component. The use of original carrier-phases and pseudo-ranges can also suppress the noise and improve the precision of real-time PPP. The results of the fault slip inversions indicate that the real-time ambiguity-fixed PPP can improve fault slip inversion and the moment magnitude estimation and become complementary to existing seismic EEW methodologies. The real-time ambiguity-fixed PPP module can be embedded into high-rate GPS receiver firmware and be incorporated into EEW systems especially for regions at threat from large magnitude earthquakes and tsunamis.

We proposed a new GPS analysis method for hazard (e.g. earthquake and tsunami) monitoring. The new augmented PPP method can overcome the limitations of current relative positioning and global PPP approaches for this application. The performance of the new approach is evaluated by GPS ground network data, observed during the 2011 Tohoku-Oki earthquake in Japan. The atmospheric corrections retrieved from the nearby monitoring stations can be interpolated with accuracy better than 5 cm. This means that the interpolated atmospheric corrections are accurate enough for rapid ambiguity resolution, which is a prerequisite to achieve the most precise displacements. The displacement waveforms, derived using the augmented PPP approach are immune to the convergence problem caused by data gaps and cycle slips and the problem of the earthquake shaking the reference station compared to the waveforms based on RP and global PPP analysis. This makes augmented PPP potentially appropriate for the application in operational earthquake/tsunami monitoring and warning systems. The reliability and accuracy of permanent coseismic displacements are also significantly improved. The RMS accuracy of about 1.4, 1.1, and 1.7 cms are achieved in the north, east, and vertical components, respectively. The inversion results indicate that the augmented PPP solution is the most consistent with post-processed ARIA solution both in the fault slip distribution and displacement fittings.

A new approach for real-time GNSS seismology using a single receiver was presented. The performance of the proposed TPP approach is validated using 1 Hz GEONET data collected during the 2011, Mw 9.0 Tohoku-Oki earthquake. When real-time precise orbit and clock corrections are available, the displacement waveforms, derived from TPP, are consistent with the post-processed PPP waveforms at an accuracy of few centimeters during the entire shaking period, even for a period of twenty minutes. The TPP permanent coseismic offsets agree with PPP ones very well with RMS values of 3.0, 2.1, and 5.6 cms in north, east, and vertical components, respectively. The results of the fault slip inversions also indicate that the TPP method can provide a reliable estimation of moment magnitude and even of the fault slip distribution. If just the broadcast orbits and clocks are available, the displacement accuracy will be degraded to some extent and this leads to underestimations of the moment magnitude and fault slip values. We compared the technical details of current single-receiver GNSS seismology approaches. Furthermore, several refinements are proposed to the variometric approach in order to eliminate the drift trend in the integrated coseismic displacements. After careful correction of all error components, the displacement waveforms derived from TPP and refined variometric approach are consistent with converged PPP at an accuracy of few centimeters. The results of the fault slip inversions also indicate that the TPP and the refined variometric approach can provide a reliable estimation of moment magnitudes and fault slip values as the converged PPP. From the above analysis and results on the single-receiver approaches for real-time GNSS seismology, we can conclude that the TPP, refined variometric approaches have equivalent mathematical model and can provide the same displacement precision with the converged PPP method. Moreover, these two approaches overcome the convergence problem of PPP, making them more suitable for seismological applications.

We presented an approach for tightly combing GPS and seismic sensor data where the accelerometer data are integrated into the ambiguity-fixed PPP processing on the observation level. The performance of the proposed tightly-integrated approach was

Chapter 6 Conclusions and Outlook

validated using the collocated high-rate GPS and strong motion data collected during the 2010, Mw 7.2 El Mayor-Cucapah earthquake and the 2011 Tohoku-Oki (Japan) earthquake. For tightly-integrated displacements, the peak displacements and long-period stability are in agreement with GPS-only solution. As a typical example, the permanent coseismic offset which is usually underestimated in the seismic-only solution can be now obtained exactly in the integrated solution. Some small-amplitude seismic details, which are not detectable in the GPS-only approach, can be detected from the tightly integrated displacements. A power spectral density analysis also demonstrates that an accurate broadband displacement waveform can be derived from the tightly-integrated filter. Furthermore, the ratio values of the tightly-integrated solutions are significantly improved from about 4 of the GPS-only solutions to 10 on average. The integrated displacements can provide the full spectrum of the seismic motion allowing the detection of P-wave arrivals and the estimation of permanent offsets. Small-scale features including P waves are visible in the integrated displacement and velocity waveforms. The P-wave arrival can be picked up accurately and used for reliable determination of epicenter and origin time. Permanent offsets can also be extracted with high accuracy and used for reliable fault slip inversion and magnitude estimation. These earthquake parameters are critical for earthquake/tsunami monitoring and early warning systems.

Up to now, 74 satellites are already in view and transmitting data compared to past years with 32 GPS-only. Once all four systems are fully deployed, about 120 navigation satellites will be available for GNSS users. Undoubtedly, the rapid development of multi-constellation GNSS could enable a wider range of more precise and reliable applications, e.g. for positioning, navigation, timing, and geophysical applications. In the near future, a joint processing of multi-GNSS (e.g., GPS, GLONASS, Galileo and BeiDou) data for better seismological applications will be investigated.

References

Allen, R., and H. Kanamori (2003). The potential for earthquake early warning in Southern California. Science, 300, 786-789.

Allen, R. & Ziv, A. (2011) Application of real-time GPS to earthquake early warning. Geophys. Res. Lett., 38, L16310.

Avallone, A., M. Marzario, A. Cirella, A. Piatanesi, A. Rovelli, C. Di Alessandro, E. D'Anastasio, N. D'Agostino, R. Giuliani, and M. Mattone (2011). Very high-rate (10 Hz) GPS seismology for moderate-magnitude earthquakes: The case of the Mw 6.3 L'Aquila (central Italy) event. J. Geophys. Res., 116(B2), B02305, doi:10.1029/2010JB007834.

Aranda, J.E., Jimenez, A., Ibarrola, G., Alcantar, F., Aguilar, A., Inostroza, M., Maldonado, S. (1995) Mexico City seismic alert system. Seismological Research Letters, 66, 42-53.

Allen, R.V. (1978) Automatic earthquake recognition and timing from single traces. B. Seismol Soc. Am., 68, 1521-1532.

Bar-Sever, Y., G. Blewitt, R. S. Gross, W. C. Hammond, V. Hsu, K. W. Hudnut, R. Khachikyan, C. W. Kreemer, R. Meyer, H.-P. Plag, M. Simons, J. Sundstrom, Y. Song, and F. Webb (2009). A GPS Real-Time Earthquake And Tsunami (GREAT) alert system, EOS Trans. AGU90, no. 52, Fall Meet. Suppl., Abstract, G21A-02.

Bassin, C., G. Laske, and G. Masters (2000). The current limits of resolution for surface wave tomography in North America.

EOS Trans., AGU 81, F897.

Blewitt, G. et al. (2009) GPS for real-time earthquake source determination and tsunami warning systems. J. Geodesy, 83, 335-343.

Blewitt, G., C. Kreemer, W. C. Hammond, and H. P. Plag (2006). Rapid determination of earthquake magnitude using GPS for tsunami warning systems. Geophys. Res. Lett., 33, L11309; doi:10.1029/GL026145.

Blewitt, G. (1989) Carrier phase ambiguity resolution for the Global Positioning System applied to geodetic baselines up to 2000 km, J. Geophys. Res., 94, 10187-10203.

Boore, D. M. (2001) Effect of baseline corrections on displacements and response spectra for several recordings of the 1999 Chi-Chi, Taiwan earthquake, Bull. Seismol. Soc. Am.

Boore, D. M., C. D. Stephens, and W. B. Joyner (2002). Comments on baseline correction of digital strong-motion data: Examples from the 1999 Hector Mine, California, earthquake, Bull. Seismol. Soc. Am., 92, 1543-1560.

Boehm, J., A. Niell, P. Tregoning, and H. Schuh (2006). Global Mapping Function (GMF): A new empirical mapping function based on numerical weather model data. Geophys. Res. Lett., 33, L7304, doi: 10.1029/2005GL025546.

Branzanti, M., G. Colosimo, M. Crespi, and A. Mazzoni (2013). GPS near-real-time coseismic displacements for the Great Tohoku-Oki Earthquake, IEEE Geosci. Remote Sens. Lett., vol. 10, no. 2, pp. 372-376.

Bock, Y., R. Nikolaidis, P. De Jonge, and M. Bevis (2000). Instantaneous geodetic positioning at medium distances with the Global Positioning System. J. Geophys. Res., 105(B12), 28,223-28,228, doi: 10.1029/JB900268.

References

Bock, Y., L. Prawirodirdjo, and T. I. Melbourne (2004). Detection of arbitrarily large dynamic ground motions with a dense high-rate GPS network. Geophys. Res. Lett., 31, doi: 10.1029/2003GL019150.

Bock, Y., D. Melgar, and B. W. Crowell (2011). Real-Time Strong-Motion Broadband Displacements from Collocated GPS and Accelerometers. Bull. Seism. Soc. Am., 101(5), doi:10.1785/0120110007.

Chan, W., Y. Xu, X. Ding, and W. Dai (2006). An integrated GPS-accelerometer data processing technique for structural deformation monitoring. J. Geodesy, 80 (12):705-719.

Collins, P., F. Lahaye, P. Hérous, and S. Bisnath (2008). Precise point positioning with AR using the decoupled clock model, In: Proceedings of IONGNSS 16-19 September, GA, USA.

Collins, P., J. Henton, Y. Mireault, P. Héroux, M. Schmidt, H. Dragert, and S. Bisnath (2009). Precise Point Positioning for Real-Time Determination of Co-seismic Crustal Motion, Proceedings of IONGNSS-2009. Savannah, Georgia, 22-25 September, pp. 2479-2488.

Crowell, B., Y. Bock, and M. Squibb (2009). Demonstration of earthquake early warning using total displacement waveforms from real-time GPS networks. Seismol. Res. Lett., 80, 772-782, doi:10.1785/gssrl.80.5.772.

Crowell, B. W., Y. Bock, and D. Melgar (2012). Real-time inversion of GPS data for finite fault modeling and rapid hazard assessment. Geophys. Res. Lett., Vol. 39, L09305, doi:10.1029/2012GL051318, 2012.

Crotwell, H. P., Owens, T. J., Ritsema, J. (1999) The Taup Toolkit: Flexible seismic travel-time and ray-path utilities. Seismological Research Letters, 70, 154-160.

Caissy, M., L. Agrotis, G. Weber, M. Hernandez-Pajares, and U. Hugentobler (2012). Coming Soon: The International GNSS Real-Time Service. GPS World, Vol. 23, Issue 6, p.52.

Caissy, M.(2006) The IGS Real-time Pilot Project-Perspective on Data and Product Generation. In: Streaming GNSS Data via Internet Symposium, 6-7, Frankfurt.

Caissy, M., L. Agrotis, G. Weber, M. Hernandez-Pajares, and U. Hugentobler (2012). Coming Soon: The International GNSS Real-Time Service. GPS World, 23(6), p. 52.

Choi, K., A. Bilich, K. Larson, and P. Axelrad (2004). Modified sidereal filtering: Implications for high-rate GPS positioning. Geophys. Res. Lett., 31, L22608, doi:10.1029/2004GL021621.

Colosimo, G., M. Crespi, and A. Mazzoni (2011). Real-time GPS seismology with a stand-alone receiver: A preliminary feasibility demonstration. J. Geophys. Res., 116, B11302, doi:10.1029/2010JB007941.

Colombelli, S., R. M. Allen, and A. Zollo (2013). Application of real-time GPS to earthquake early warning in subduction and strike-slip environments. J. Geophys. Res. Solid Earth, 118 (7), 3448-3461, doi:10.1002/jgrb.50242.

Dong, D. N. & Bock, Y. (1989) Global Positioning System network analysis with phase ambiguity resolution applied to crustal deformation studies in California. J. Geophys. Res., 94, 3949-3966.

Diao, F., X. Xiong, R. Wang, Y. Zheng, and H. Hsu (2010). Slip model of the 2008 Mw 7.9 Wenchuan (China) earthquake derived from the co-seismic GPS data. Earth, Planets and Space, 62, 869-874.

Dow, J. M., R. E. Neilan, and C. Rizos (2009). The International GNSS Service in a changing landscape of Global Navigation Satellite Systems. J. Geod., 83(7), 191-198, doi:10.1007/

s00190-008-0300-3.

Emore, G.L., Haase, J.S., Choi, K., Larson, K.M., Yamagiwa, A. (2007) Recovering seismic displacements through combined use of 1-Hz GPS and strong-motion accelerometers. B. Seismol Soc. Am., 97, 357-378.

Espinosa-Aranda, J., A. Jimenez, G. Ibarrola, F. Alcantar, A. Aguilar, M. Inostroza, and S. Maldonado (1995). Mexico City seismic alert system. Seismol. Res. Lett., 66 (6), 42-53.

Ge, M., G. Gendt, M. Rothacher, C. Shi, and J. Liu (2008). Resolution of GPS carrier-phase ambiguities in precise point positioning (PPP) with daily observations. J. Geod, 82(7): 389-399. doi:10.1007/s00190-007-0187-4.

Ge, M., J. Dousa, X. Li, M. Ramatschi, and J. Wickert (2011). A Novel Real-Time Precise Positioning Service System: Global Precise Point Positioning with Regional Augmentation, In: Proceedings of the 3rd Int. Colloquium-Galileo Science, 31 August-2 September 2011, Copenhagen, Denmark.

Ge, M., Dousa, J., Li, X., Ramatschi, M. & Wickert, J. (2012) A Novel Real-Time Precise Positioning Service System: Global Precise Point Positioning with Regional Augmentation. JGPS, 11, 2-10.

Ge, L., (1999). GPS seismometer and its signal extraction, in Proceedings of the 12th International Technical Meeting of the Satellite Division of the Institute of Navigation. Institute of Navigation, Fairfax, VA, pp. 41-52.

Ge, L., S. Han, C. Rizos, Y. Ishikawa, M. Hoshiba, Y. Yoshida, M. Izawa, N. Hashimoto, and S. Himori (2000). GPS Seismometers with up to 20 Hz Sampling Rate. Earth, Planets, Space, 52(10), 881-884.

Gao, Y.; Shen, X. (2001) Improving ambiguity convergence in carrier phase-based precise point positioning. In Proceedings of the 14th International Technical Meeting of the Satellite Division of the Institute of Navigation (ION GPS 2001). Salt Lake City, UT, USA, pp. 1532-1539.

Geng, J., X. Meng, A. Dodson, M. Ge, and F. Teferle (2010). Rapid re-convergences to ambiguity-fixed solutions in precise point positioning. J. Geodesy, Doi:10.1007/s00190-010-0404-4.

Geng, J., C. Shi, M. Ge, A. H. Dodson, Y. Lou, Q. Zhao, and J. Liu (2012). Improving the estimation of fractional-cycle biases for ambiguity resolution in precise point positioning. J. Geod., 86, 579-589, doi:10.1007/s00190-011-0537-0.

Geng, J., Y. Bock, D. Melgar, B. W. Crowell, and J. Haase (2013). A new seismogeodetic approach applied to GPS and accelerometer observations of the 2012 Brawley seismic swarm: implications for earthquake early warning. Geochem. Geophys. Geosyst., 14(7), 2124-2142, doi: 10.1002/ggge.20144.

Genrich, J. F., and Y. Bock (2006). Instantaneous geodetic positioning with 10-50 Hz GPS measurements: Noise characteristics and implications for monitoring networks. J. Geophys. Res., 111, doi 10.1029/2005JB003617.

Hoechner, A., M. Ge, A. Y. Babeyko, and S.V. Sobolev (2013). Instant tsunami early warning based on real-time GPS-Tohoku 2011 case study. Nat. Hazards Earth Syst. Sci., 13, 1285-1292, doi:10.5194/nhess-13-1285-2013.

Han, S. (1997) Carrier phase-based long-range GPS kinematic positioning. PhD thesis, School of Geomatic Engineering, The University of New South Wales.

Han, S. (1997) Quality-control issues relating to instantaneous ambiguity resolution for real-time GPS kinematic positioning. J Geodesy, 71, 351-361.

Hirahara, K., Nakano, T. & Hoso, Y. (1994) An experiment for

References

GPS strain seismometer. in Proc. of the Japanese Symposium on GPS. 15-16 December, Tokyo, Japan, 67-75.

Harris, R., and P. Segall (1987). Detection of a locked zone at depth on the Parkfield, California segment of the San Andreas Fault. J. Geophys. Res., 92, 7945-7962.

Jackson, M. (2003) Geophysics at the speed of light: Earth Scope and the Plate Boundary Observatory, the Leading Edge, 22, 262-267, doi:10.1190/1.1564532.

Kogan, M.G., W.Y. Kim, Y. Bock, and A.W. Smyth (2008). Load response on the Verrazano Narrows Bridge during the NYC Marathon revealed by GPS and accelerometers. Seismol. Res. Lett., 79, 12-19.

Kanamori, H. (2007) Real-time earthquake damage mitigation measures. In: Earthquake Early Warning Systems. Eds. P. Gasparini, G. Manfredi, and J. Zschau, 1-8. Berlin and Heidelberg: Springer (ISBN-13 978-3-540-72240-3).

Kouba, J, and Héroux P. (2001) Precise point positioning using IGS orbit and clock products. GPS Solutions, 5(2): 12-28. doi:10.1007/PL00012883.

Kouba, J. (2003) Measuring seismic waves induced by large earthquakes with GPS. Stud. Geophys. Geod., 47, 741755.

Li, X., and X. Zhang (2012). Improving the Estimation of Uncalibrated Fractional Phase Offsets for PPP Ambiguity Resolution. J. Navig., 65, pp. 513-529 doi:10.1017/S0373463 312000112.

Li, X. (2012) Improving Real-time PPP Ambiguity Resolution with Ionospheric Characteristic Consideration. Proc. of ION GNSS-12, Institute of Navigation, Nashville, Tennessee, September, 17-21.

Li, X., X. Zhang, and M. Ge (2011), Regional reference network augmented precise point positioning for instantaneous ambiguity resolution. J. Geodesy, 85, 151-158.

Li, X., M. Ge, X. Zhang, Y. Zhang, B. Guo, R. Wang, J. Klotz, and J. Wickert (2013a). Real-time high-rate co-seismic displacement from ambiguity-fixed precise point positioning: Application to earthquake early warning. Geophys. Res. Lett., 40(2), 295-300, doi:10.1002/grl.50138.

Li, X., M. Ge, B. Guo, J. Wickert, and H. Schuh (2013b). Temporal point positioning approach for real-time GNSS seismology using a single receiver. Geophys. Res. Lett., 40 (21), 5677-5682, doi:10.1002/2013GL057818.

Li, X., M. Ge, H. Zhang, and J. Wickert (2013c). A method for improving uncalibrated phase delay estimation and ambiguity-fixing in real-time precise point positioning. J. Geod., 87(5), 405-416, doi:10.1007/s00190-013-0611-x.

Li, X., M. Ge, Y. Zhang, R. Wang, P. Xu, J. Wickert, and H. Schuh (2013d). New approach for earthquake/tsunami monitoring using dense GPS networks. Sci. Rep., 3, 2682, doi:10.1038/srep02682

Li, T., and J. Wang (2012). Some Remarks on GNSS Integer Ambiguity Validation Methods. Survey Review., 44 (326): 320-328.

Loyer, S., F. Perosanz, F. Mercier, H. Capdevilleet, and J. Marty (2012). Zero-difference GPS ambiguity resolution at CNES-CLS IGS Analysis Center. J. Geodesy, 2012, doi: 10.1007/s00190-012-0559-2

Larson, K., P. Bodin, and J. Gomberg (2003). Using 1-Hz GPS data to measure deformations caused by the Denali fault earthquake. Science, 300, 1421-1424.

Larson, K. (2009) GPS seismology. J. Geodesy, 83, 227-233, doi: 10.1007/s00190-008-0233-x.

Laurichesse, D., F. Mercier, J.P. Berthias, and J. Bijac (2008).

Real-time zero-difference ambiguities fixing and absolute RTK, In: Proceedings of ION National Technical Meeting. San Diego, CA, US.

Lee, V. W., and M. D. Trifunac (2009). Empirical scaling of rotational spectra of strong earthquake ground motion. Bull. Seismol. Soc. Am., 99, 1378-1390.

Loyer, S., F. Perosanz, F. Mercier, H. Capdevilleet, and J. Marty (2012), Zero-difference GPS ambiguity resolution at CNES-CLS IGS Analysis Center. J. Geod., 2012, doi: 10.1007/s00190-012-0559-2.

Meng, X. (2013) From Structural Health Monitoring to Geo-Hazard Early Warning: An Integrated Approach Using GNSS Positioning Technology. In Earth Observation of Global Changes (EOGC), Springer: pp. 285-293.

Melgar, D., Y. Bock, and B. W. Crowell (2012). Real-Time Centroid Moment Tensor Determination for Large Earthquakes from Local and Regional Displacement Records. Geophys. J. Int., 188, 703-718, doi: 10.1111/j.1365-246X.2011.05297.x.

Melgar, D., Y. Bock, D. Sanchez, and B. W. Crowell (2013). On robust and reliable automated baseline corrections for strong motion seismology. J. Geophys. Res. Solid Earth, 118, 1177-1187, doi:10.1002/jgrb.50135.

Nikolaidis, R., Y. Bock, P. J. de Jonge, P. Shearer, D. C. Agnew, and M. Van Domselaar (2001). Seismic wave observations with the Global Positioning System. J. Geophys. Res., 106, 21,897-21,916.

Ohta, Y., T. Kobayashi, H. Tsushima, S. Miura, R. Hino, T. Takasu, H. Fujimoto, T. Iinuma, K. Tachibana, and T. Demachi (2012). Quasi real-time fault model estimation for near-field tsunami forecasting based on RTK-GPS analysis: Application to the 2011 Tohoku-Oki earthquake (Mw 9.0). J.

Geophys. Res., 117 (B2), B02,311.

Picozzi, M.; Bindi, D.; Pittore, M.; Kieling, K.; Parolai, S. (2013) Real-time risk assessment in seismic early warning and rapid response: a feasibility study in Bishkek (Kyrgyzstan). J. Seismol, 17, 485-505.

Remondi, B., (1985) Performing centimeter-level surveys in seconds with GPS carrier phase: initial results. Navigation, 32:386-400.

Schaffrin, B., and Y. Bock (1988). A unified scheme for processing GPS dual-band phase observations. Bull. Geod., 62 pp. 142-160.

Smyth, A., and M. Wu (2006). Multi-rate Kalman filtering for the data fusion of displacement and acceleration response measurements in dynamic system monitoring. Mech. Syst. Signal Process, 21, 706-723.

Schaer, S., G. Beutler, M. Rothacher, E. Brockmann, A. Wiger, and U. Wild (1999). The impact of the atmosphere and other systematic errors on permanent GPS networks. Proc. IAG Symposium on Positioning, Birmingham, UK, 19-24 July, 406.

Simons, M., S. Minson, A. Sladen, F. Ortega, J. Jiang, S. Owen, L. Meng, J. Ampuero, S. Wei, R. Chu, D. Helmberger, H. Kanamori, E. Hetland, A. Moore, F. Webb (2011). The 2011 magnitude 9.0 Tohoku-Oki earthquake: Mosaicking the megathrust from seconds to centuries. Science, 322, 1426, doi:10.1126/science.120702.

Saastamoinen, J. (1972) Atmospheric correction for the troposphere and stratosphere in radio ranging satellites, in The Use of Artificial Satellites for Geodesy, Geophys. Monogr. Ser., Vol. 15, edited by S. W. Henriksen, A. Mancini, and B. H. Chovitz, pp. 247-251, AGU, Washington, D. C., doi: 10.1029/GM015p0247.

Teunissen, P. J. G. (1995) The least squares ambiguity decorrelation adjustment: a method for fast GPS integer estimation. J. Geodesy, 70, 65-82.

Teunissen, P. J. G., and A. Kleusberg (1996). GPS for Geodesy. Volume 60 of Lecture Notes in Earth Sciences, pp. 175-217.

Trifunac, M. D., and M. I. Todorovska (2001). A note on the usable dynamicrange of accelerographs recording translation. Soil Dyn. Earthq. Eng., 21, 275-286.

Trnkoczy, A., Havskov, J., Ottem O Ller, L. (2002) Seismic networks. In New Manual of Seismological Observatory Practice 2 (NMSOP-2), Bormann, P., Ed. Deutsches Geo. Forschungs Zentrum GFZ: Potsdam, Germany, Vol. 2, pp. 1-65.

Tu, R., M. Ge, R. Wang, and T. R. Walter (2014). A new algorithm for tight integration of real-time GPS and strong-motion records, demonstrated on simulated, experimental, and real seismic data. J. Seismol., 18, no. 1, 151-161.

Wang, R., Lorenzo-Martín, F. & Roth, F. (2003). Computation of deformation induced by earthquakes in a multi-layered elastic crust—FORTRAN programs EDGRN/EDCMP. Comput. Geosci, 29, 195-207.

Wang, R., B. Schurr, C. Milkereit, Zh. Shao, and M. Jin (2011). An improved automatic scheme for empirical baseline correction of digital strong-motion records. Bull. Seism. Soc. Am., 101(5), 2029-2044, doi: 10.1785/0120110039.

Wang, R., S. Parolai, M. Ge, M. Ji, T.R. Walter, and J. Zschau (2013). The 2011 Mw 9.0 Tohoku-Oki Earthquake: Comparison of GPS and Strong-Motion Data. Bull. Seism. Soc. Am., 103 (2b), doi: 10.1785/0120110264.

Wessel, P. & Smith, W. H. F. (1998) New, improved version of generic mapping tools released. EOS Trans. AGU, 79, 579-

579.

Wright, T. J., N. Houlié, M. Hildyard, and T. Iwabuchi (2012). Real-time, reliable magnitudes for large earthquakes from 1 Hz GPS precise point positioning: The 2011 Tohoku-Oki (Japan) earthquake. Geophys. Res. Lett., 39, L12302, doi: 10.1029/2012GL051894.

Xu, C., Y. Liu, Y. Wen and R. Wang (2010). Coseismic Slip Distribution of the 2008 Mw 7.9 Wenchuan Earthquake from Joint Inversion of GPS and in SAR Data. Bull. Seism. Soc. Am., 100(5B), 2736-2749, doi: 10.1785/0120090253.

Xu, P., C. Shi, and J. Liu (2012). Integer estimation methods for GPS ambiguity resolution: an applications oriented review and improvement. Survey Review, 44:59-71.

Xu, P., Shi, C., Fang, R., Liu, J., Niu, X., Zhang, Q., Yanagidani, T. (2013) High-rate precise point positioning (PPP) to measure seismic wave motions: an experimental comparison of GPS PPP with inertial measurement units. J Geodesy, 87, 361-372.

Yang, Y., He, H. & Xu, G. (2001) Adaptively robust filtering for kinematic geodetic positioning. J. Geodesy, 75, 109-116.

Yokota, Y., K. Koketsu, K. Hikima, and S. Miyazaki (2009). Ability of 1-Hz GPS data to infer the source process of a medium-sized earthquake: The case of the 2008 Iwate-Miyagi Nairiku, Japan, earthquake, Geophys. Res. Lett., 36(12), L12301, doi:10.1029/2009GL037799.

Zhang, X., and X. Li (2012). Instantaneous Re-initialization in Real-time Kinematic PPP with Cycle-slips Fixing. GPS Solutions, 16(3), pp.315-327, doi: 10.1007/s10291-011-0233-9.

Zhang, X., X. Li, and F. Guo (2011). Satellite Clock Estimation at 1 Hz for Realtime Kinematic PPP applications. GPS Solutions, Vol. 15, No. 4, 315-324, doi: 10.1007/s10291-010-

References

0191-7.

Zhang, X., and B. Guo (2013). Real-time tracking the instantaneous movement of crust during earthquake with a stand-alone GPS receiver. Chinese J. Geophys.(in Chinese), 56(6), 1,928-1,936, doi: 10.6038/cjg20130615.

Zumberge, J.F., M.B. Heflin, D.C. Jefferson, M.M. Watkins, and F.H. Webb (1997). Precise point positioning for the efficient and robust analysis of GPS data from large networks. J. Geophys. Res., 102(B3): 5005-5017. doi: 10.1029/96JB03860. 91, 1199-1211.

Acknowledgments

The completion of this book would not have been possible without the support of many people and organizations.

First of all, I would like to express my profound respect and gratitude to my supervisors, Prof. Harald Schuh at TU Berlin and GFZ, Dr. Maorong Ge, and Prof. Urs Hugentobler, for supervising this book, for their continuous support, guidance and encouragement of my studies and research. They were always willing to share their insights with me and were patient with every question.

I am grateful to Prof. Dr. Harald Schuh, director of GFZ Department 1 for his kindly help, organization of my work and improvements of my publications. I also would like to thank Dr. Jens Wickert for his helpful suggestions for my publications and PhD work.

I owe a special thank to my supervisor Dr. Maorong Ge, whom I thank so much for his continuous and extensive support throughout the last years. His selfless help, detailed guidance and discussion are the source of my forward momentum. I would also like to thank my GFZ colleagues Dr. Dick Galina, Dr. Mathias Fritsche, Dr. Florian Zus, Dr. Rongjiang Wang, Dr. Yong Zhang, Dr. Jurgen Klotz, Dr. Tong Ning, Dr. Tobias Nilsson, Dr. Robert Heinkelmann, Dr. Jan Dousa, Dr. Zhiguo Deng, Dr. Michal Bender, Dr. Gerd Gendt, Dr. Markus Ramatschi, Dr. Faqi Diao, Mr. Maik Uhlemann, Mr. Markus Bradke, Mr. Andre Brandt, Mr.

Acknowledgments

Thomas Nischan, Dr. Junping Chen, Dr. Ming Shangguan, Dr. Rui Tu, Xiaolei Dai, Hua Chen, Kaifei He, Kejie Chen, and others, for their generous help and discussions. I appreciate their kindly help, cooperation and discussions. I have enjoyed working with all of my colleagues.

I am indebted to my host institute GFZ where I was doing my research from 2010 to 2015. TU Berlin is also thanked for my study during this period. The International GNSS Service (IGS) are thanked for providing GNSS data and products.

Finally, I would like to express my gratitude to my parents for their endless support and understanding. The persons I wish to thank most are my wife and daughter, who fill my heart with confidence, courage, and love all the time.